51 High-Tech Practical Jokes for the Evil Genius

Evil Genius Series

Bionics for the Evil Genius: 25 Build-It-Yourself Projects

Electronic Circuits for the Evil Genius: 57 Lessons with Projects

Electronic Gadgets for the Evil Genius: 28 Build-It-Yourself Projects

Electronic Games for the Evil Genius

Electronic Sensors for the Evil Genius: 54 Electrifying Projects

50 Awesome Auto Projects for the Evil Genius

50 Model Rocket Projects for the Evil Genius

51 High-Tech Practical Jokes for the Evil Genius

Fuel Cell Projects for the Evil Genius

Mechatronics for the Evil Genius: 25 Build-It-Yourself Projects

MORE Electronic Gadgets for the Evil Genius: 40 NEW Build-It-Yourself Projects

101 Outer Space Projects for the Evil Genius

101 Spy Gadgets for the Evil Genius

123 PIC® Microcontroller Experiments for the Evil Genius

123 Robotics Experiments for the Evil Genius

PC Mods for the Evil Genius

Radio and Receiver Projects for the Evil Genius

Solar Energy Projects for the Evil Genius

25 Home Automation Projects for the Evil Genius

51 High-Tech Practical Jokes for the Evil Genius

BRAD GRAHAM

KATHY McGOWAN

New York   Chicago   San Francisco   Lisbon   London   Madrid
Mexico City   Milan   New Delhi   San Juan   Seoul
Singapore   Sydney   Toronto

1 2 3 4 5 6 7 8 9 0 QPD/QPD 0 1 3 2 1 0 9 8 7

ISBN 978-0-07-149494-6
MHID 0-07-149494-4

Sponsoring Editor: Judy Bass
Production Supervisor: Pamela A. Pelton
Editing Supervisor: Stephen M. Smith
Project Manager: Imran Mirza
Copy Editor: Clare Freeman
Proofreader: Ian Ross
Indexer: Walter Schneider
Art Director, Cover: Jeff Weeks
Composition: Keyword Group Ltd.

Printed and bound by Quebecor/Dubuque.

McGraw-Hill books are available at special quantity discounts to use as premiums and sales promotions, or for use in corporate training programs. For more information, please write to the Director of Special Sales, McGraw-Hill Professional, Two Penn Plaza, New York, NY 10121-2298. Or contact your local bookstore.

This book is printed on acid-free paper.

"I don't know about that Graham boy"—*Concerned parent*

About the Authors

Brad Graham is an inventor, robotics hobbyist, founder and host of the ATOMICZOMBIE.COM web site (which receives over 2.5 million hits monthly), and a computer professional. He is the co-author, with Kathy McGowan, of *101 Spy Gadgets for the Evil Genius*, *Atomic Zombie's Bicycle Builder's Bonanza* (perhaps the most creative bicycle-building guide ever written), and *Build Your Own All-Terrain Robot*, all from McGraw-Hill. Technical manager of a high-tech firm that specializes in computer network setup and maintenance, data storage and recovery, and security services, Mr. Graham is also a Certified Netware Engineer, a Microsoft Certified Professional, and a Certified Electronics and Cabling Technician.

Kathy McGowan provides administrative, logistical, and marketing support for Atomic Zombie's™ many robotics, bicycle, technical, and publishing projects. She also manages the daily operations of a high-tech firm and several web sites, including ATOMICZOMBIE.COM, as well as various Internet-based blogs and forums. Additionally, Ms. McGowan writes articles for e-zines and is collaborating with Mr. Graham on several film and television projects.

Acknowledgments

Our Evil Genius collaborator Judy Bass at McGraw-Hill has always been our biggest fan and we can't thank her enough for believing in us every step of the way. A heartfelt thank you to Judy and everyone at McGraw-Hill for helping to make this project a reality. Thanks also to all of you who contact us, especially members of the "Atomic Zombie Krew," our international family of Evil Geniuses, bike builders, and robotics junkies. We sincerely appreciate your support, friendship, and feedback. You're the best creative "krew" in the world.

There are many projects, a blog, videos, a builder's gallery, and support at ATOMICZOMBIE.COM. We always look forward to seeing what other Evil Geniuses are up to. Hope to see you there!

Cool stuff, cool people!

ATOMICZOMBIE.COM

Contents

1 Introduction   **1**
   Warranty Void!   1
   Basic Electronics   2

2 Truly Annoying Devices   **17**
   Project 1—The Dripping Faucet   17
   Project 2—Evasive Beeping Thing   20
   Project 3—Ghost Door Knocker   24
   Project 4—Putrid Stink Machine   28

3 Critters and Beasties   **33**
   Project 5—Alive and Breathing   33
   Project 6—Hairy Swinging Spider   36
   Project 7—Carpet Crawling   40
      Creature
   Project 8—Universal Critter   44
      Launcher
   Project 9—Trash Can Troll   48

4 Mechanical Mayhem   **51**
   Project 10—Remote Control   51
      Jammer
   Project 11—Radio Station Blocker   54
   Project 12—Video Fubarizer   57
   Project 13—Audio Distorter   59
   Project 14—Phone Static Injector   61
   Project 15—Hard Drive Failure   63
   Project 16—Serious Car   68
      Troubles

5 Things That Go Bump in the Night   **71**
   Project 17—Glowing Blinking Eyes   71
   Project 18—Computer Audio   76
      Nightmare
   Project 19—Rats in the Walls   82
   Project 20—Footsteps in the Night   84
   Project 21—Giant Shadow   86
      Projector

6 Evil Abounds!   **91**
   Project 22—Voices from the Grave   91
   Project 23—Evil-Possessed Doll   95
   Project 24—Telephone Devil Voice   98
   Project 25—Evil Lurching Head   101
   Project 26—Give Us a Sign!   104
   Project 27—Flying Ouija Board   107

7 Shock and Awe!   **113**
   Project 28—The Barbeque Box   113
   Project 29—Simple Induction Shocker   115
   Project 30—Strong Pulse Shocker   117
   Project 31—Disposable Camera Zapper   120
   Project 32—Hissing Gas Container   124
   Project 33—Radiation Detector   127

8 Machine Hoaxes   **133**
   Project 34—The Magic Light Bulb   133
   Project 35—Coin-Minting Machine   136
   Project 36—See Through Walls   142

9 Mind Benders   **147**
   Project 37—Rigged Lie Detector   147
   Project 38—The Dog Talker   149
   Project 39—Telepathy Transmitter   153
   Project 40—Subliminal Audio   157
      Mind Control
   Project 41—Subliminal Video   160
      Mind Control

10 Halloween Horrors   **165**
   Project 42—Flying Vampire Bat   165
   Project 43—The Haunted Ghost Mirror   170
   Project 44—Living Brain in a Jar   172
   Project 45—Universal Motivator   176
   Project 46—Sound-activated Switch   179
   Project 47—Flesh-eating   181
      Jack-O-Lantern

11 Fluffy Attacks! Scare Them Silly!   **187**

    Project 48—Spring-Loaded   188
           Launch Pad

    Project 49—Trap Door Cage   190

    Project 50—Light-activated Trigger   193

    Project 51—Fluffy's Body   198

12 Digital Fakery   **203**

    Editing Software   203

    Original Photo Quality   204

    Warping Effects   204

    Making Hoax Photos   206

    Adding Reflections and Shadows   212

Index   **221**

Contents

51 High-Tech Practical Jokes for the Evil Genius

Chapter 1

Introduction

Warranty void!

This book was written for all those who feel the irresistible urge to break open the case to see what makes that appliance or electronic device work. "There are no user serviceable parts inside," or "disassembly will void the warranty" are phrases that simply fuel the fire for us hardware-hacking Evil Geniuses. The ability to make an electronic or mechanical device do things that it was not intended for is a skill that is easily learned by anyone who is not afraid to put his or her crazy ideas to the test, and possibly blow a few fuses or fry a few circuits along the way. You do not need an engineering degree or a room full of sophisticated tools to become a successful hardware hacker, just the desire to create, a good imagination and a large pile of junk to experiment with.

A warped sense of humor can be a venerable force when mixed with the ability to turn evil mechanical ideas into real-world working devices. I believe that if you are planning to do something, you should make it count. As all of my once-unsuspecting friends can attest to, this attitude applies to my practical jokes as well. Of course, you must remember the "golden rule," and expect that your practical joke victims will some day turn the tables on you. You never know who might have a copy of this book, and a list with your name on it! Of course, all of the evil ideas in this book are designed to be harmless, even though some of them may be quite elaborate in nature. Knowing when not to launch a prank, and learning to weed out those who have no sense of humor is also a skill that should be practiced, and you will have a great time with the projects in this book.

If you have never cracked the case on an electronic device, or have never wielded the unlimited power of the almighty soldering iron, then fear not—I have not used any rare parts or special tools, just hardware store parts, common appliances and basic tools. To gain the most from this book, don't be afraid to alter the projects to suit your needs. You can mix and match different projects to create thousands of new devices to perform your evil bidding. This is hacking after all, and it would be unbecoming of an Evil Genius to fully follow the instructions. Another thing you may notice that is missing from this book is a rigid parts list. Rather than specifying a "50-megawatt ruby laser" (only available from a particular website or store), I have tried to use only the most common parts found by butchering standard easy-to-find appliances or parts found off the shelf from any hardware store. Also, many of the parts can be substituted for similar parts that will do the same job and, as you get better at hacking and inventing, you will be able to turn just about any pile of junk into something wonderful. This way, you can work with what you have available without breaking your budget in the process, or spending weeks waiting for some overpriced exotic part to arrive in the mail from afar.

For those who are just starting a career as an Evil Genius hardware hacker, take your time and don't give up if things don't turn out the way you expected on the first try. Hey, we all have to start at the beginning, and thanks to the Internet, you should be able to find the answers you seek very easily. There are hundreds of in-depth tutorials that can help you understand basic concepts that may not be familiar to you, such as LED theory, using transistors, or just basic polarity and electrical theory. You may consider joining a few electronic forums on the Internet, as there is a wealth of knowledge, and many experienced members who may be willing to answer your questions. If you are a "newbie," don't let that fact discourage you from seeking answers; even the brightest electronic engineers could not identify the positive terminal on a capacitor at one point in their early careers.

Well, that pretty much sums up my introduction. Just take your time, feel free to experiment, and don't be afraid to put your ideas into motion! The basic electronics theory that follows covers most of the technology used in this book, and can be used to create just about any electronic device imaginable, since many large circuits are nothing more than many smaller simpler circuits working together.

Basic electronics

Electronics is the art of controlling the electron, and semiconductors are the tools that make this possible. "Semiconductor" is the name given to the vast quantity of various components used to generate, transform, resist and control the flow of electrons in order to achieve some goal. If you have ever had the chance to look at a large main board from a device such as a computer or video player, then you would have seen the vast city of semiconductors interconnected by thousands of tiny wires scattered around the circuit board that holds them all in place. At first glance, this intricate city of complexity may be overwhelming and impossible to understand, but in reality, all of these semiconductors do a very basic task by themselves, and these tasks are not hard to understand once you know the basics. Even a very complex integrated circuit with hundreds of tiny pins, such as a 1 million gate FPGA, is nothing more than a collection of smaller semiconductors such as resistors and transistors densely packed into a microscopic area using state of the art manufacturing processes. Having an understanding of the most basic electronic building blocks will

allow you to understand even the most complex designs. I am not going to dig as far down as atomic theory or how the various components are manufactured since that would double the size of this book and bore you to tears. I will, however, cover each of the most basic semiconductors that form the building block of many larger circuits as well as the tools and techniques that you will need to work with them. If you want to dig deeper into electronics theory, then find a nice thick book loaded with formulas or spend some time on the Internet researching the areas that may interest you—the wealth of knowledge on the Internet regarding electronics and hardware hacking in general is as far reaching as the ends of the galaxy! Now, let's start by covering the mandatory tools and techniques you will need for this hobby.

Basic tools

If you do not already have a soldering iron, then drop this book and head down to your local hobby or electronics store and get one because you will not be able to build even the most basic circuit without one. Of course, like any tool of the trade, you can get a basic model for a few bucks, or go for the deluxe model with all the bells and whistles such as digital heat control, ergonomic grip and who knows what else. The soldering iron shown in Figure 1-1 would be considered medium quality, and it comes with a holster and basic heat control.

I will admit that I have never owned anything more than a $10 black handle soldering iron and have built some very small circuit boards using surface-mounted components without any real problem. I am not saying that you shouldn't spend the money for a quality soldering station, it is indeed worth it, but not absolutely necessary to get started. To feed your soldering iron, you will need a roll of "flux" core solder, which is probably the only type you will find at most hobby or electronics supply outlets. Flux is a reducing agent designed to help remove impurities (specifically oxidized metals) from the points of contact to improve the electrical connection between the semiconductor lead and the copper traces on a circuit board. Flux core solder is manufactured as a hollow tube and filled with the flux so that it is applied as you melt the solder. Solder used for electronics work is not the same as the heavy solid type used for plumbing, which is meant to be applied with a torch or high-heat soldering gun. The solder you will need will only be a millimeter in diameter and probably come on a small spool or coiled up in a plastic tube with a label that reads something like 40/60, indicating the percentage of

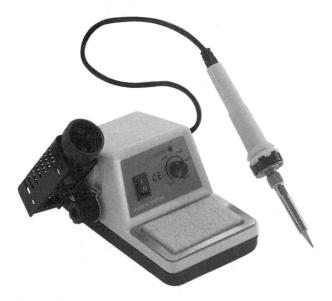

Figure 1-1 *Soldering iron with heat control*

tin and lead in the solder. With a decent soldering iron and a roll of flux core solder, you will be able to remove and salvage semiconductors from old circuit boards or create your own circuits from scratch using pre-drilled copper-plated boards or by simply soldering the leads together with wires. There is one more soldering tool which I find to be a lifesaver, especially if you do a lot of circuit design and do not like waiting for days for some oddball value semiconductor to arrive in the mail. This tool, shown in Figure 1-2, is a spring-activated vacuum and is commonly called a "solder sucker."

When you are salvaging components from old circuit boards, it can be very difficult to extract the ones that have more than a few leads by simply heating up the solder side of the board as you pull on the component, so you will have to find a way to extract the solder from each lead to free the component. The solder sucker does a marvelous job of removing the molten solder by simply pressing down on the lever once the spring has been loaded to create a vacuum, which draws the molten solder into the tube and away from the circuit board and component leads. Using this simple heat and suck process, you can remove parts with many leads, such as large integrated circuits, with great speed and ease, and without

Figure 1-2 *A solder sucker tool*

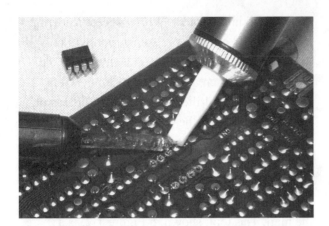

Figure 1-3 *Removing an integrated circuit with the solder sucker tool*

Figure 1-4 *A basic multi-meter for electronics work*

much risk of overheating the component or fine copper traces. Figure 1-3 shows the solder sucker removing the solder from the last leg of an 8-pin op amp of some defunct DVD player main board. When you build up a nice stock of circuit boards, you will save a ton of time and money when you want a part that would normally have to be ordered.

Considering a typical DVD player or VCR main board could have 500 resistors, 100 capacitors, 50 transistors and diodes, and hundreds of other useful components, this handy solder sucker can turn a discarded electronic appliance into hundreds of dollars worth of semiconductors, so collect as many old circuit boards as you have room for. Most of the semiconductors used for the various projects in this book came from old circuit boards, and it is not very often that I have to order new parts unless working on a cutting-edge design or something really non-standard.

Now, there is one last tool you will need to have in your electronics toolkit, and this is a multi-meter, which can measure voltage, resistance, and possibly capacitance and frequency. It's pretty hard to troubleshoot a failing circuit without some kind of voltage test, and you will certainly need to measure impedance when checking the values of semiconductors such as resistors, coils, transistors and diodes. Even the most basic and inexpensive

multi-meter will have these functions. Of course, you can find a lot more in a desktop multi-meter, and it usually boils down to how much you are willing to spend vs. what you really need. I have a basic hardware-store variety digital multi-meter (Figure 1-4) that can measure AC and DC voltage, amperage, resistance, capacitance and frequencies up to 10 MHz. This unit is considered entry level, and does the job for 90 percent of all the analog and digital projects that I tinker with. When I really get deep into the high-speed circuitry such as radiofrequency devices or high-speed microcontrollers, I find myself using an oscilloscope to examine microsecond timings and extremely weak analog signals, but for basic electronic circuits such as those presented in this book, an oscilloscope will not be necessary.

So there you have it—with a soldering iron, a roll of solder, a solder sucker, a basic multi-meter, and a pile of old circuit boards, you can build just about anything you want as long as you have the basic know how and patience. Now, let's have a look at what the most common semiconductors do, and learn how to identify them.

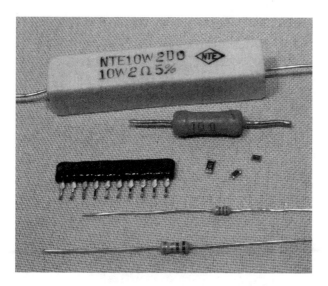

Figure 1-5   *Several typical resistors*

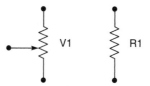

Figure 1-6   *Resistor schematic symbols*

Resistors

Resistors, like the ones shown in Figure 1-5, are the most basic of the semiconductors you will be using, and they do exactly what their name implies—they resist the flow of current by exchanging some current for heat, which is dissipated through the body of the device. On a large circuit board, you could find hundreds of resistors populating the board, and even on tiny circuit boards with many surface-mounted components, resistors will usually make up the bulk of the semiconductors. The size of the resistor generally determines how much heat it can dissipate and will be rated in watts, with ¼ and ⅛ watts being the most common type you will work with (the two bottom resistors in Figure 1-5). Resistors can become very large, and will require ceramic-based bodies, especially if they are rated for several watts or more, like the 10-watt unit shown at the top of Figure 1-5.

Because of the recent drive to make electronics more "green" and power-conservative, large, power-wasting resistors are not all that common in consumer electronics these days, since it is more efficient to convert amperage and voltage using some type of switching power supply or regulator rather than by letting a fat resistor burn away the energy as heat. On the other hand, small-value resistors are very common, and you will find yourself dealing with them all of the time for simple tasks such as driving an LED with limited current, pulling up an input pin to a logical "one" state, biasing a simple transistor amplifier, and thousands of other common functions. On most common axial lead resistors, like the ones you will most often use in your projects, the value of the resistor is coded onto the device in the form of four colored bands which tell you the resistance in "ohms." Ohms are represented using the Greek omega symbol (Ω), and will often be omitted for values over 99 ohms, which will be stated as 1K, 15K, 47K, or some other number followed by the letter K, indicating the value is in kilo ohms (thousands of ohms). Similarly, for values over 999K, the letter M will be used to show that 1M is actually 1 mega ohm, or one million ohms. In a schematic diagram, a resistor is represented by a zigzag line segment as shown in Figure 1-6, and will either have a letter and a number such as R1 or V3 relating to a parts list, or will simply have the value printed next to it such as 1M, or 220 ohms. The schematic symbol on the left of Figure 1-6 represents a variable resistor, which can be set from zero ohms to the full value printed on the body of the variable resistor.

A variable resistor is also known as a "potentiometer," or "pot," and it can take the form of a small circuit-board mounted cylinder with a slot for a screwdriver, or as a cabinet-mounted can with a shaft exiting the can for mating with some type of knob or dial. When you crank up the volume on an amplifier with a knob, you are turning a potentiometer. Variable resistors are great for testing a new design, since you can just turn the dial until the circuit performs as you want it to, then remove the variable resistor to measure the impedance (resistance) across the leads in order to determine the best value of fixed resistor to install. On a variable resistor, there are usually three leads: the outer two connect to the fixed carbon resistor inside the can, which gives the variable resistor its value, and a center pin that connects to a wiper, allowing the selection of resistance from zero to full. Several common variable resistors are shown in Figure 1-7, with the top left unit dissected to show the resistor band and wiper.

As mentioned earlier, most resistors will have four color bands painted around their bodies, which can be decoded into a value as shown in Table 1-1. At first, this may seem a bit illogical, but once you get the hang of the color

Figure 1-7 *Common variable resistors*

Table 1-1
Resistor color chart

Color	1st Band	2nd Band	3rd Band	Multiplier
Black	0	0	0	
Brown	1	1	1	1 Ω
Red	2	2	2	10 Ω
Orange	3	3	3	100 Ω
Yellow	4	4	4	1 KΩ
Green	5	5	5	10 KΩ
Blue	6	6	6	100 KΩ
Violet	7	7	7	1 MΩ
Gray	8	8	8	10 MΩ
White	9	9	9	0.1
Gold				0.01
Silver				

band decoding, you will be able to recognize most common values at first glance without having to refer to the chart.

There will almost always be either a silver or gold band included on each resistor, and this will indicate the end of the color sequence, and will not become part of the value. A gold band indicates the resistor has a 5 percent tolerance (margin of error) in the value, so a 10K resistor could end up being anywhere from 9.5K to 10.5K in value, although in most cases will be very accurate. A silver band indicates the tolerance is only 10 percent, but I have yet to see a resistor with a silver band that was not on a circuit board that included vacuum tubes, so forget that there is even such a band! Once you ignore the gold band, you are left with three color bands that can be used to determine the exact value as given in Table 1-1. So let's say we have a resistor with the color bands brown, black, red, and gold. We know that the gold band is the tolerance band and the first three will indicate the values to reference in the chart. Doing so, we get 1 (brown), 0 (black), and 100 ohms (red). The third band is the multiplier, which would indicate that the number of zeros following the first two values will be 2, or the value is simply multiplied by 100 ohms. This translates to a value of 1000 ohms, or 1K (10 × 100 ohms). A 370K resistor would have the colors orange, violet, and yellow followed by a gold band. You can check the value of the resistor when it is not connected to a circuit by simply placing your multi-meter on the appropriate resistance scale and reading back the value. I do not want to get too deep into electronics formulas and theory here, since there are many good books dedicated to the subject, so I will simply leave you with two basic rules regarding the use of resistors: put them in series to add their values together, and put them in parallel to divide them. This simple rule works great if you are in desperate need of a 20K resistor, for instance, but can only find two 10K resistors to put in series. In parallel, they will divide down to 5K. Now you can identify the most common

semiconductor that is used in electronics today, the resistor, so we will move ahead to the next most common semiconductor, the capacitor.

Capacitors

A capacitor in its most basic form is a small rechargeable battery with a very short charge and discharge cycle. Where a typical AAA battery may be able to power an LED for a month, a capacitor of similar size will power it for only a few seconds before its energy is fully discharged. Because capacitors can store energy for a predictable duration, they can perform all kinds of useful functions in a circuit, such as filtering AC waves, creating accurate delays, removing impurities from a noise signal, and creating clock and audio oscillators. Because a capacitor is basically a battery, many of the large ones available look much like batteries with two leads connected to one side of a metal can. As shown in Figure 1-8, there are many sizes and shapes of capacitors, some of which look like small batteries.

Just like resistors, capacitors can be as large as a coffee can, or as small as a grain of rice, it really depends on the value. The larger devices can store a lot more energy. Unlike batteries, some capacitors are non-polarized, and they can be inserted into a circuit regardless of current flow, while some cannot. The two different types of capacitors are shown by their schematic symbols

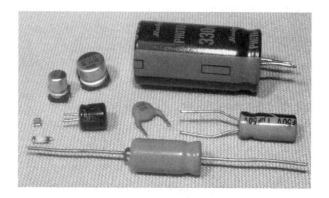

Figure 1-8   *Various common capacitors*

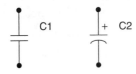

Figure 1-9 *Capacitor symbols*

in Figure 1-9, C1 being a non-polarized type, and C2 a polarized type. Although there are always exceptions to the rules, generally the disk-style capacitors are non-polarized, and the larger can-style electrolytic types are polarized. An obvious indicator of a polarized capacitor is the negative markings on the can, which can be clearly seen in the larger capacitor shown at the top of Figure 1-8.

Another thing that capacitors have in common with batteries is that polarity is very important when inserting polarized capacitors into a circuit. If you install an electrolytic capacitor in reverse and attempt to charge it, the part will most likely heat up and release the oil contained inside the case causing a circuit malfunction or dead short. In the past, electrolytic capacitors did not have a pressure release system, and would explode like firecrackers when overcharged or installed in reverse, leaving behind a huge mess of oily paper and a smell that was tough to forget. On many capacitors, especially the larger can style, the voltage rating and capacitance value is simply stamped on the case. A capacitor is rated in voltage and in farads, which defines the capacitance of a dielectric for which a potential difference of one volt results in a static charge of one coulomb. This may not make a lot of sense until you start messing around with electronics, but you will soon understand that typically, the larger the capacitor, the larger the farad rating will be, thus the more energy it can store. Since a farad is quite a large value, most capacitors are rated in microfarads (μF), such as the typical value of 4700 μF for a large electrolytic filter capacitor, and 0.1 μF for a small ceramic disk capacitor. Picofarads (pF) are also used to indicate very small values such as those found in many ceramic capacitors or adjustable capacitors used in radiofrequency

circuits (a pF is one millionth of a μF). On most can-style electrolytic capacitors, the value is simply written on the case and will be stated in microfarads and voltage along with a clear indication of which lead is negative. Voltage and polarity are very important in electrolytic capacitors, and they should always be inserted correctly, with a voltage rating higher than necessary for your circuit. Ceramic capacitors will usually only have the value stamped on them if they are in picofarads for some reason, and often no symbol will follow the number, just the value. Normally, ceramic capacitors will have a three-digit number that needs to be decoded into the actual value, and this evil scheme works as shown in Table 1-2.

Who knows why they just don't write the value on the capacitor? I mean, it would have the same amount of digits as the code! Oh well, you get used to seeing these codes, just like resistor color bands, and in no time will easily recognize the common values such as 104, which would indicate a 0.1 μF value according to the chart. Capacitors behave just like batteries when it comes to parallel and series connections, so, in parallel, two identical capacitors will handle the same voltage as a single unit, but double their capacitance rating, and in series they have the same capacitance rating as a single unit, but can handle twice the voltage. So if you need to filter a really noisy power supply, you might want to install a pair of 4700 μF capacitors in parallel to end up with a capacitance of 9400 μF. When installing parallel capacitors, make sure that the voltage rating of all the capacitors used are higher than the voltage of that circuit, or there will be a failure.

Diodes

Diodes allow current to flow through them in one direction only so they can be used to rectify AC into DC, block unwanted current from entering a device, protect a circuit from a power reversal, and even give off light in the case of light-emitting

Table 1-2

Ceramic capacitor value chart

Value	Marking	Value	Marking
10 pf	10 or 100	0.0033 µF	332
12 pf	12 or 120	0.0039 µF	392
15 pf	15 or 150	0.0047 µF	472
18 pf	18 or 180	0.0056 µF	562
22 pf	22 or 220	0.0068 µF	682
27 pf	27 or 270	0.0082 µF	822
33 pf	33 or 330	0.01 µF	103
39 pf	39 or 390	0.012 µF	123
47 pf	47 or 470	0.015 µF	153
58 pf	58 or 580	0.018 µF	183
68 pf	68 or 680	0.022 µF	223
82 pf	82 or 820	0.027 µF	273
100 pf	101	0.033 µF	333
120 pf	121	0.039 µF	393
150 pf	151	0.047 µF	473
180 pf	181	0.056 µF	563
220 pf	221	0.068 µF	683
270 pf	271	0.082 µF	823
330 pf	331	0.10 µF	104
390 pf	391	0.12 µF	124
470 pf	471	0.15 µf	154
560 pf	561	0.18 µF	184
680 pf	681	0.22 µF	224
820 pf	821	0.27 µF	274
0.001 µF	102	0.33 µF	334
0.0012 µF	122	0.39 µF	394
(1200 pf)		0.47 µF	474
0.0015 µF	152	0.56 µF	564
0.0018 µF	182	0.68 µF	684
(1800 pf)		0.82 µF	824
0.0022 µF	222	1 µF	105 or 1µF
0.0027 µF	272		

diodes (LEDs). Figure 1-10 shows various sizes and type of diodes including an easily recognizable LED and the large full-wave rectifier module at the top. A full-wave rectifier is just a block containing four large diodes inside.

Like most other semiconductors, the size of the diode is usually a good indication of how much

current it can handle before failure, and this information will be specified by the manufacturer by referencing whatever code is printed on the diode to some data sheet. Unlike resistors and capacitors, there is no common mode of identifying a diode unless you get to know some of the most common manufacturers' codes by

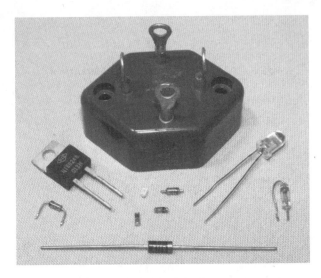

Figure 1-10 *Several styles of diodes*

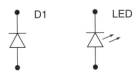

Figure 1-11 *Diode schematic symbols*

memory, so you will be forced to look up the data sheet on the Internet or in a cross-reference catalog to determine the exact value and purpose of unknown diodes. For example, the NTE6248 diode shown in Figure 1-10 in the TO220 case (left side of photo) has a data sheet that indicates it is a Schottky barrier rectifier with a peak reverse-voltage maximum of 600 volts and a maximum forward current rating of 16 amps. Data sheets will tell you everything you need to know about a particular device, and you should never exceed any of the recommended values if you want a reliable circuit. The schematic symbol for a diode is shown in Figure 1-11, D1 being a standard diode, and the other a light-emitting diode (the two arrows represent light leaving the device).

The diode symbol shows an arrow (anode) pointing at a line (cathode), and this will indicate which way current flows (from the anode to the cathode, or in the direction of the arrow). On many small diodes, there will be a stripe painted around the case to indicate which end is the cathode,

and on LEDs, there will be a flat side on the case nearest the cathode lead. LEDs come in many different sizes, shapes, and wavelengths (colors), and have ratings that must not be exceeded in order to avoid damaging the device. Reverse voltage and peak forward current are very important values that must not be exceeded when powering LEDs or damage will easily occur, yet at the same time, you will want to get as close as possible to the maximum values if your circuit demands full performance from the LED, so read the data sheets on the device carefully. Larger diodes used to rectify AC or control large current may need to be mounted to the proper heat sink in order to operate at their rated values, and often the case style will be a clear indication due to the metal backing or mounting hardware that may come with the device. Unless you know how much heat a certain device can dissipate in open air, your best bet is to mount it to a heat sink if it was designed to be installed that way. Like most semiconductors, there are thousands of various sizes and types of diodes, so make sure you are using a part rated for your circuit, and double check the polarity of the device before you turn on the power for the first time.

Transistors

A transistor is one of the most useful semiconductors available, and often the building block for many larger integrated circuits and components such as logic gates, memory and microprocessors. Before transistors became widely used in electronics, simple devices like radios and amplifiers would need huge wooden cabinets, consume vast amounts of power, and emit large wasteful quantities of heat due to the use of vacuum tubes. A vacuum-tube-based computer called ENIAC was once built that used 17,468 vacuum tubes, 7,200 crystal diodes, 1,500 relays, 70,000 resistors, 10,000 capacitors and had more than 5 million hand-soldered joints. It weighed 30 tons and was roughly 8 feet × 3 feet × 100 feet, and consumed 150 kW of power! A simple computer that would rival the power of this power hungry monster could easily be built on a few square inches of perforated board using a few dollars in parts today by any electronics hobbyist, thanks to the transistor. A transistor is really just a switch that can control a large amount of current by switching a small amount of current, thus creating an amplifier. Several common types and sizes of transistors are shown in Figure 1-12.

Depending on how much current a transistor is designed to switch, it may be as small as a grain of rice or as large as a hockey puck and require a massive steel heat sink or fan to operate correctly. There are thousands of varying transistor types and sizes, but one thing most of them have in common is that they will have three connections that can be called "collector," "emitter" and "base," and will be represented by one of the two schematic symbols shown in Figure 1-13.

Figure 1-12   *Various common transistors*

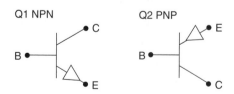

Figure 1-13   *NPN and PNP transistor schematic symbols*

The emitter (E), base (B) and collector (C) on both the negative-positive-negative (NPN) and positive-negative-positive (PNP) transistors do the same job. The collector/emitter current is controlled by the current flowing between the base and emitter terminals, but the flow of current is opposite in each device. Today, most transistors are NPN due to the fact that it is easier to manufacture a better NPN transistor than a PNP, but there are still occurrences when a circuit may use a PNP transistor due to the direction of current, or in tandem with an NPN transistor to create a matched pair. There is enough transistor theory to cover ten books of this size, so I will condense that information in order to help you understand the very basics of transistor operation. As a simple switch, a transistor can be thought of as a relay with no mechanical parts. You can turn on a high-current load such as a light or motor with a very weak current such as the output from a logic gate or light-sensitive photocell. Switching a large load with a small load is very important in electronics, and transistors do this perfectly and at speed that a mechanical switch such as a relay could never come close to achieving. A audio amplifier is nothing more than a very fast switch that takes a very small current such as the output from a CD player and uses it as the input into a fast switch that controls a large current such as the DC power source feeding the speakers. Almost any transistor can easily operate well beyond the frequency of an audio signal, so they are perfectly suited for this job. At much higher frequencies like those used in radio transmitters, transistors do the same job of amplification, but are rated for much higher frequencies sometimes into the gigahertz range. Another main difference between the way a mechanical switch and a transistor work is the fact that a transistor is not simply an on or off switch, it can operate as an "analog" switch, varying the amount of current switched by varying the amount of current entering the base of the transistor. A relay can turn on a 100-watt light bulb if a 5-volt current is applied to the coil, but a transistor could vary the intensity of the same light bulb from zero to full brightness depending on the voltage at the base.

Amplifier Transistors

NPN Silicon

MAXIMUM RATINGS

Rating	Symbol	Value	Unit
Collector-Emitter Voltage	V_{CEO}	40	Vdc
Collector-Base Voltage	V_{CBO}	75	Vdc
Emitter-Base Voltage	V_{EBO}	6.0	Vdc
Collector Current—Continuous	I_C	600	mAdc
Total Device Dissipation @ T_A = 25°C Derate above 25°C	P_D	625 5.0	mW mW/°C
Total Device Dissipation @ T_C = 25°C Derate above 25°C	P_D	1.5 12	Watts mW/°C
Operating and Storage Juntion Temperature Range	T_J,T_{stg}	−55 to +150	°C

THERMAL CHARACTERISTICS

Characteristic	Symbol	Max	Unit
Thermal Resistance, Junction to Ambient	$R_{\theta JA}$	200	°C/W
Thermal Resistance, Junction to Case	$R_{\theta JC}$	83.3	°C/W

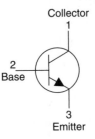

P2N2222A

1
2
3

CASE 29–11, STYLE 17
TO–92 (TO–226AA)

Collector
1

2
Base

3
Emitter

Figure 1-14   *Data sheet for the common 2N2222 NPN transistor*

Like all semiconductors, the transistor must be rated for the job you intend it to do, so maximum current, switching voltage and speed are factors that need to be considered when choosing the correct part. The data sheet for a very common NPN transistor, the 2N2222 (which can be substituted for the 2N3904 often used in this book) is shown in Figure 1-14.

From this page, we can see that this transistor can switch about half a watt (624 mW) with a voltage of 6 volts across the base and emitter junction. Of course, these are maximum ratings, so you might decide that the transistor will work safely in a circuit if it had to switch on a 120-mW LED from a 5-volt logic level input at the base. As a general rule, I would look at the maximum switching current of a transistor, and never ask it to handle more than half of the rated maximum value, especially if it was the type of transistor designed to be mounted to a heat sink. The same thing applies to maximum switching

speed—don't expect a 100-MHz transistor to oscillate at 440 MHz in an RF transmitter circuit, since it will have a difficult enough time just reaching the 100-MHz level.

Breadboards and circuit boards

Once you find a project and the parts needed to build it, you will need to connect all the leads from each semiconductor together in order to create the completed circuit. A commercial product will have a printed circuit board, perfectly made with one or more layers, and could contain thousands of semiconductors of all sizes including surface-mounted devices, each with hundreds of pins per package. A circuit board of this magnitude is well out of reach for the average hobby builder, so unless you want to spend a few hundred dollars to have a single circuit board made, you will need to find another way to get those

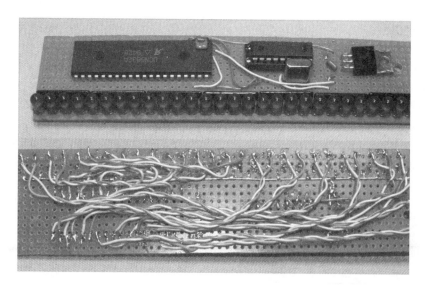

Figure 1-15 *Perforated board is great for making circuits*

semiconductors connected. Sure, you could send your design to one of those fast turnaround printed circuit board manufacturers that charge under a hundred dollars for a few boards, but what if you decide to change something, or realize one of the parts you planned to use is now in a different package layout? The best way to build a single circuit board is by simply hand wiring it to a bit of perforated board, especially if the parts count is low and there are now extremely high frequencies in use. Every project in this book that has a schematic diagram was built by placing the semiconductor leads through the holes on a bit of perforated board, and then soldering the underside using either the leads of each component, or a bit of wire. Figure 1-15 shows one of my "perf. board" projects built by dropping all the semiconductors on the board and wiring them on the underside. This device includes a microprocessor with custom software that magically draws an image in mid-air using 32 pulsed LEDs as you wave the unit back and forth like a flag. If you want to know how a device like this works, search Google for "scanned LED", or visit www.atomiczombie.com and check out LED scanner in the electronics projects section of our gallery.

This circuit may seem to be very complex, especially with all that wiring on the underside of the board, but in reality, it is a very simply circuit, and all those wires connect the 32 LEDs to the LED driver chip. In the early days of computer design, entire 8-bit computer systems were built using this same technique, although they had thousands of wires. If there is a problem with a part, or some of the wiring, then just get out your soldering tools and fix it. The same easy repair would not happen on a printed circuit board, which can become a real problem for those who mass produce electronic devices. This perforated board can be purchased at any electronics supply or hobby shop in squares ranging in size from a few inches to a foot or more, and you can just snap off as much as you need for whatever circuit you plan to build. There are also prototyping circuit boards available that have solder pads on one or both sides so you can make a more permanent circuit board with a lot less hand wiring by connecting the pads together with solder. An example of this type of prototyping board is shown in Figure 1-16 with the components for one of my robot stepper motor drives on it.

You can also get "proto boards" with solder pads connected in rows: a very easy design, with minimal wiring, as well as specially shaped blank cards for designing and testing circuits that might need to plug into a computer slot, or satellite dish. The fact is, there are plenty of ways to build most simple projects without having to spend a ton of money and time trying to acquire a professionally made printed circuit board. You could even go as far as etching your own copper boards into printed circuit boards, and there are numerous Internet resources that will show you how to do this, or sell complete kits based on chemical etching, or photo etching of copper plates. I rarely use the etching technique since it will have the same downfalls of

the professionally made circuit board when it comes to easy modification, and require a lot more work. If you are designing a circuit from scratch, or want to try a schematic found on the Internet, it may be a lot of work to solder all the components on a perf. board or proto board just to see if the thing even works, so many hobbyists start without any soldering at all using what is called a "breadboard." A typical breadboard will have many strips of interconnected electrical terminals, known as "bus strips," down one or both sides, either as part of the main unit or as separate blocks clipped on to carry the power rails. This allows you to simply press the semiconductor leads into the breadboard and interconnect the circuit using small bits of wire with the insulation removed at each end. To make changes, simply move the wiring around. Figure 1-17 shows the solderless breadboard I used to develop all the circuits in this book and many hundreds of other projects.

Also shown in Figure 1-17 (inset) is the relationship between the connected strips and the holes in the board. The Atmel processor plugged into the top of the board would have every leg connected to a vertical strip of five interconnected

Figure 1-16   *A solder pad prototyping board*

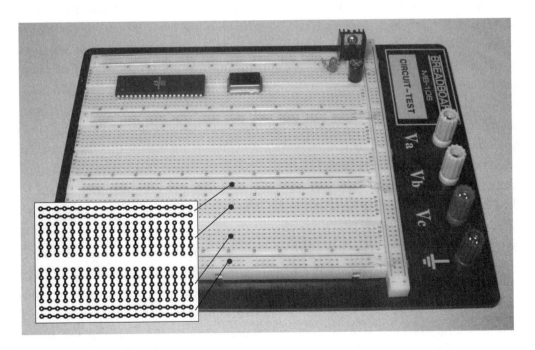

Figure 1-17   *A solderless breadboard*

Introduction

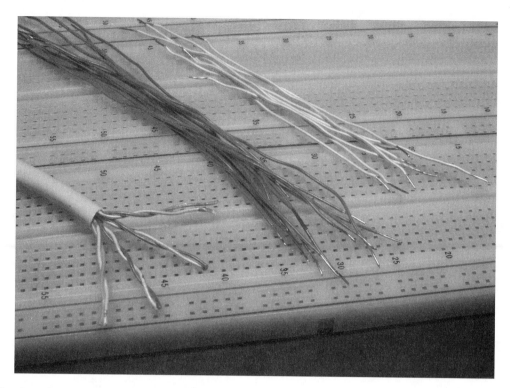

Figure 1-18   *Breadboard wiring*

holes, so you can place a wire in any of the four open holes and make a connection to the pin on that strip. Breadboards are a hobbyist's best friend, and I certainly recommend that you purchase one or more of them if you plan to make anything more complex than an LED flasher circuit. I have learned one important thing after owning several different models of breadboards—purchase a quality unit with a metal base or your high-speed circuits will fail. If you have a microprocessor clocked over 4 MHz, or any RF circuit on a breadboard, it will act glitchy on a cheap breadboard with no metal base due to stray and unpredictable capacitance. I have run processors over 40 MHz on my breadboard and designed working RF transmitters in the 500-MHz range with few problems other than a slight re-tuning after moving them to a real circuit board. Another tip that will save you a lot of messing around is that the perfect wire for these breadboards can be found by cutting up some CAT-5 network cable as shown in Figure 1-18. This solid core copper wire is inexpensive, easy to strip, color coded and

works perfectly in all breadboards that I have used over the years.

It is a good idea to cut up many various lengths of breadboarding wire ahead of time so you can concentrate on designing your circuit. Try to avoid stranded wire as well, since it will be difficult to insert into the holes and may tend to bunch up and make a faulty connection, which could be a real problem to track down.

Well, that's it! With a handful of semiconductors, a breadboard and soldering iron, you should be able to create just about anything you like. Don't give up every time blue smoke pours out of transistor, or when a circuit does something completely unexpected, it's all part of this game. Learn as you go, using the Internet, reference books, and other people's designs as a guide and, before long, you will be able to whip up any type of circuit without any reference material at all.

Let's start building some truly annoying devices to help you hone your electronics skills.

Chapter 2

Truly Annoying Devices

Project 1—The Dripping Faucet

This device makes the sound of dripping water. It's very difficult to find because it only makes a sound in complete darkness. The unit is very sensitive to any amount of light, so even the faint glow from a nearby night light will make it go silent, causing great frustration for the person who is trying to find the source of the leak. Build the unit into a small plastic box, or conceal it in a familiar kitchen or bathroom object, such as a cup or tissue box, to make finding the device even harder. The dripper will run for days on a single 9-volt battery, and will certainly drive anyone mad trying to locate it.

The unit is made using two common 555 timers—one that sends out a series of timed pulses when the lights are out, and another that makes a high-pitched chirp each time the pulse is sent. The rate of repetition can be set from several seconds to a few times per second so that the unit can be made to sound like a slow leak or a type of insect. The chirp frequency can be adjusted as well to tailor the sound to both the container used to hold the device and the type of "piezo element" used to produce the sound.

A piezo element is nothing more than a bit of piezoceramic material glued to a metal disk so that it will resonate when a current is applied. A piezo element by itself cannot generate any sound, which is why the second 555 timer is used as an audio oscillator. Piezo elements are easy to find at any electronics supplier, and in many electronic appliances such as microwave ovens,

cash registers, computers, digital watches (the back cover is the piezo element), and practically any device that makes a beep or blip sound. A piezo element is easy to identify, and it may come in several varieties as shown in Figure 2-1.

As shown in Figure 2-1, plastic-encased piezo elements come in sizes from less than an inch in diameter to several inches. The unit on the top right of the picture is the bare element, which can sometimes be found glued directly to the cabinets of some electronic devices such as telephones, toys or even the backs of digital-watch covers. Sometimes it is easy to confuse an encased piezo element with an audio buzzer, since they often look the same. An audio buzzer is designed to make a sound as soon as power is applied, and because it already contains an audio oscillator, will often have a voltage rating or pin polarity stamped on the case. If you are not sure, just apply 5 or 9 volts (take note of the polarity if it is indicated on the case), and listen for a sound. A piezo element will only make a single pop, whereas a buzzer will produce a sound. Piezo elements are not polarity sensitive, so it does not matter which pin is positive or negative. Have a look at the schematic for the dripper as shown in Figure 2-2. I will explain how it works and how you can alter it to make different sounds.

As stated earlier in this section, there are two 555 timers used. The function of timer 1 is to create a series of pulses that vary between several

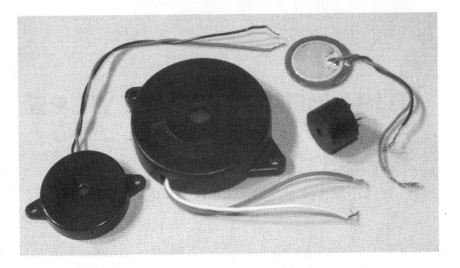

Figure 2-1 *Several piezo elements*

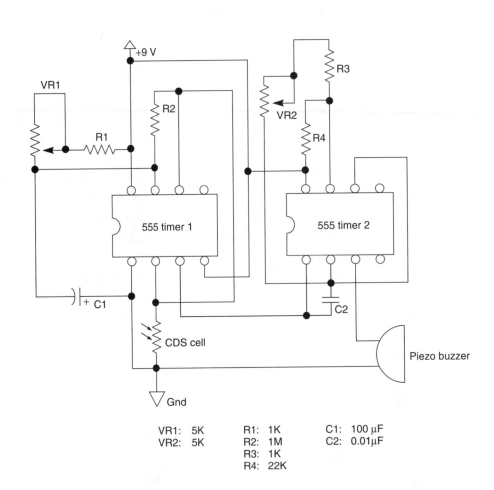

VR1:	5K	R1:	1K	C1:	100 µF
VR2:	5K	R2:	1M	C2:	0.01µF
		R3:	1K		
		R4:	22K		

Figure 2-2 *Dripping faucet simulator schematic*

seconds each and several pulses per second. The pulse rate is controlled by setting variable resistor VR1 to the desired rate. Timer 1 will only begin to send pulses if there is no light in the area, since it is controlled by the CDS cell shown on pin 2, which will almost short the pin to ground when any light strikes its surface. When there is no light present, the CDS cell reaches a megaohm or more and the timer can resume its job of sending out pulses on pin 3, which feeds the second timer. The second timer is a basic audio oscillator that can be set to various high frequencies by adjusting variable resistor VR2. The output of timer 2 is sent directly to the piezo element on pin 3 to produce a very short duration high-pitched noise that sounds a lot like a water drop or a pest. If you want to play around with more varying sound frequencies and timing rates, then you can mess around with the values of VR1, VR2, R1 and R3 to make some very interesting sounds. Another thing that can alter the sound is the type of enclosure used. Sound waves will resonate differently depending on both the shape and material used to make the enclosure.

The dripper can be built on a bit of perforated board and hand wired as shown in Figure 2-3. It will run for many days connected to a good 9-volt battery, but can run from as low as 5 volts, and as high as 12 volts without a problem.

Figure 2-3 shows the completed circuit ready to be engaged for hours of great fun at the expense of someone's good night sleep! The circuit is simple enough to hand wire on the underside of the perforated board using some hookup wire and a soldering iron. Then, it was tested with a fresh battery and the lights off. If you find that the unit will not start when the lights are out, test first to ensure that there is sound output by removing the CDS cell completely, which will cause the unit to start dripping. When you reinstall the CDS, the unit will stay quiet until there is absolutely no light at all in the room. Even the smallest bit of light will silence the device, so if your target room has any ambient light, you will have to add a resistor in series with the CDS cell in order to make the unit less sensitive to ambient light. Try a 50–100K resistor or a variable 100K resistor in series with the CDS cell to help the unit switch on in a dimly lit environment. A little experimentation may be necessary, but once working, the dripper will spring into action as soon as the light is off and become silent as soon as someone turns on the light to investigate.

Figure 2-3 *The dripper circuit assembled*

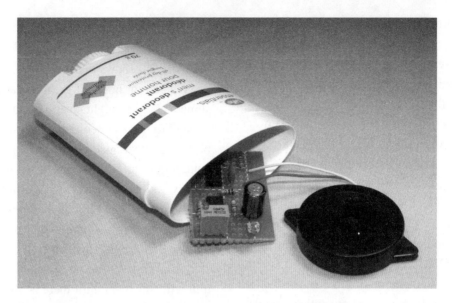

Figure 2-4 *The dripper circuit assembled*

Where to hide such a device? Well, there are so many possibilities. I found that an empty deodorant case worked perfectly as shown in Figure 2-4. There was enough room for the battery, the piezo element, and the circuit board. Owing to the extreme sensitivity of the CDS cell and slightly opaque plastic, I did not even have to drill a hole for the CDS cell. The sound was also loud enough that a hole for the piezo element was not needed, especially since my covert container could exist in direct sight without detection. When you are cramming all of the guts into the enclosure, be careful that the underside of the circuit board does not touch anything conductive like the piezo metal or battery casing. Wrapping the circuit board in a bit of paper towel or fastening all the parts to the plastic may be a good idea, especially if you plan to move the unit around.

Figure 2-5 shows the completed dripping faucet device ready to ruin a good night's sleep when the light goes out. This unit can even detect the hallway light, so it really is difficult to track down unless you have night vision goggles! In Figure 2-5, you can see the small switch that I added in series with one of the battery connections to turn the device off when it is not in use. If you want to expand the vocabulary of this device, you could add two cabinet mounted variable resistors for VR1 and VR2 to allow it to be turned into a cricket, mouse, fast drip, or just about any similar sound by simply turning the two knobs. Speed up the chirp, and hide the unit in a tent to simulate an insect infestation, or hide it in the trash can. Yes, the ways you can annoy your friends with this simple device are truly endless!

Project 2—Evasive Beeping Thing

The evasive beeping thing is appropriately named, since it dutifully does exactly what its name implies: it sends out a 5-second high-pitched beep every few minutes. The source is extremely difficult to locate because of the way that high frequencies can penetrate objects and trick our ears. You have probably encountered something similar in the real world such as a failing appliance, noisy

Figure 2-5   *This annoying device eludes detection*

video screen, or even a beeping wrist watch buried deep in a couch. As you know, high-pitched sounds seem like they are coming from all directions, which makes tracking them to the source a real chore. Add the fact that the sound only happens once every several minutes, and it may drive a person loopy as they spend all day looking for the source of the sound. Well, that's our goal anyhow!

To generate the high pitch audio wave, a small speaker like those found in tiny electronic devices (cell phones, transistor radios or a tweeter from a small speaker system, etc.) will be connected to a simple audio oscillator set to a frequency near the upper limits of our audio capabilities. The oscillator is triggered to run for approximately five seconds every few minutes by a 555 timer circuit with its output connected to the oscillator. The higher the frequency rating of the speaker, the farther the high-pitched sound will travel, which is why a two- or three-inch diameter tweeter is optimal for this project. The small speakers shown in Figure 2-6 are perfect for this project, and I included a piezo buzzer as described in the last project, as it can also be used with a simple modification of the oscillator circuit.

The rating of the small speaker is not important, since the audio oscillator will drive speakers from

4 to 16 ohm with very little power output. The speaker on the bottom left was the loudest of the ones that I tested since it was an actual tweeter removed from a small boom-box cabinet, and strictly designed to pass high frequencies. The speaker shown on the top left was the one I decided to use in the final design though because it fit nicely into the cabinet I chose to help disguise the evil device. Now, let's get on to the design of the electronics that make this unit work.

Figure 2-7 shows the schematic of the beeping thing, and you may recognize some similarities between this schematic and the last one, as they are based on similar parts and principals. In the schematic, the 555 is set up so that its output will turn on the two transistor audio oscillators formed by the pair of NPN transistors. Just like most 555 timer circuits, the timing cycle is controlled by the two resistors on pins 6, 7 and 8, and by the capacitor connected to pins 1 and 2. If you play around with the values of the two resistors, you can control the duty cycle of the timing pulses in order to alter both the off time and on time of the cycle to create more or less beep each time the cycle repeats. The capacitor controls the actual frequency of the timing pulses, and the larger the value, the longer the duration between each

Figure 2-6 *Several small high-frequency speakers*

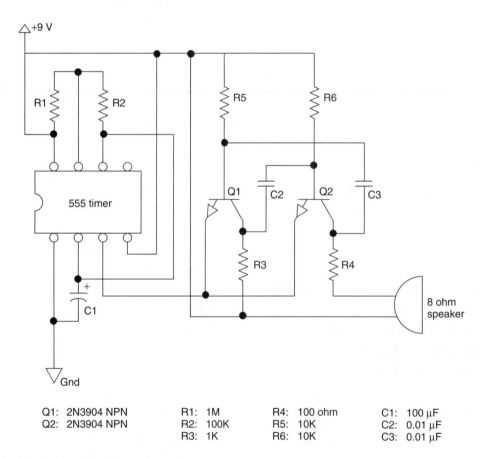

Q1: 2N3904 NPN	R1: 1M	R4: 100 ohm	C1: 100 µF
Q2: 2N3904 NPN	R2: 100K	R5: 10K	C2: 0.01 µF
	R3: 1K	R6: 10K	C3: 0.01 µF

Figure 2-7 *Evasive beeping-thing schematic*

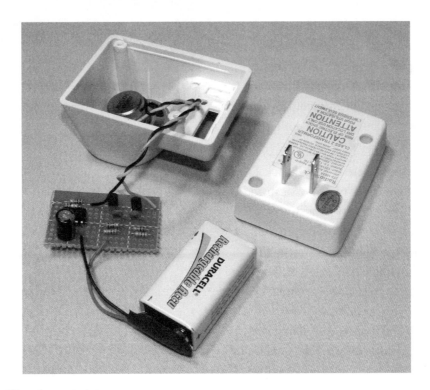

Figure 2-8   *Installing the parts into a case*

timing cycle. In a really large room, you might want a longer beep and a longer cycle, so a 220 µF capacitor could be used, and the 100K resistor could be swapped for a 220K resistor. For a smaller room, where it may be easier to locate the device (e.g. a friend's office), the capacitor could be changed to 47 µF and the 100K resistor to a 10K for a very short beep. The best plan is to simply build the unit as is and then fine tune the components until you are happy with its operation. And yes, a variable resistor would be easy to adjust.

Now, where do you hide the beast? Well, since this unit emits hard-to-locate high frequencies, your options are endless. The high-pitched sound will exit through the smallest hole in whatever box you place the parts into. I decided to cram the works into an old wall adapter that has all of the guts removed, including any connection to the AC lines. The little speaker fits nicely into the top of the box, and there was just enough room for the 9-volt battery and small circuit board. Figure 2-8 shows the completed circuit going into the wall wart box.

There was just enough room to get all the parts inside, so I could not install an on-off switch, but

that was OK since the top of the box simply snapped together and I could simply unclip the battery. The unit will run for many days on a full battery, and if you strategically place the beeper, it may take that long for the unsuspecting victim to find it! If you plan to use a wall wart cabinet for the device like I did, ensure that there is no connection between the plug prongs and the AC lines. It is a good idea to remove them completely. Some other good hiding places might be a pop can, lunch box, wall clock, tissue box, or you could even install it into a working appliance. A solid cabinet will need a small hole for the speaker to optimize the distance that the sound will travel. I found that a quarter-inch hole was large enough for the tiny two-inch speaker I used. As I mentioned earlier, you can also use a piezo buzzer instead of a speaker, which would make the unit even smaller and possibly louder owing to the very good high-pitched operation of the piezo element. To use a piezo buzzer in place of the speaker, connect resistor R4 (which used to connect to one of the speaker terminals) directly to the +9 line, where the other speaker terminal used to connect.

Now you can place the piezo buzzer in parallel with R4 to make it function. The reason this is done is because the piezo element will offer very high resistance as compared to the very low resistance of the speaker, and the current from the battery needs to flow to transistor Q2's collector.

The final product shown in Figure 2-9 looks at home just about anywhere there is a wall socket, and can be easily hidden under furniture or inside another appliance for truly covert mind-warping annoying fun and games. I have covered up the voltage switch from the original wall wart with black tape, and the little hole on the top of the case is barely large enough to pass a decent amount of high-pitched sound. With the component values given, the beep emits about once every three minutes and lasts for approximately five seconds, just enough time to entice the victim to look for the source of the sound before it goes silent. I like to drop the unit in a room, then claim that I can't hear any beeping. This really gets the "beeper hunter" ticked off, and they try even harder to track down the evasive beeping thing to no avail.

Figure 2-9   *What is that annoying beeping sound?*

"I don't hear anything pal, maybe you need an ear exam, or you should stop listening to pirated music on your MP3 player. I heard that the new copy protection can make your ears ring for days!"

Project 3—Ghost Door Knocker

This interesting project combines a little hardware with some timing electronics to simulate the sound of someone knocking on a door or wall. To create a circuit that sounds like three or four quick knocks, a 555 timer is used as a long delay counter set to stay off for a few minutes and then send out a pulse for about three seconds. This three-second pulse is not much use by itself, so it is fed into a PNP transistor in order to switch on a double-pole, double-throw relay, which is configured in such a way that it turns itself on and off several times per second. The other relay pole is then used to bang a washing machine or photocopier solenoid plunger up and down against the enclosure or wall to simulate the sound of rapping at the door. The operation of the timing hardware will be explained in more detail later, so dig into your scrap bin or head down to the surplus electronics store and try to find a 5- or 12-volt mechanical solenoid like the ones shown in Figure 2-10.

A mechanical solenoid is nothing more than an electromagnet with a steel plunger placed in the center of the coil so that, when energized, the plunger will pull itself into the coil as far as it can go. Figure 2-10 shows a pair of solenoids taken from a photocopier and a washing machine with one of the plungers removed from the electromagnet. The solenoid on the left is rated for 12 volts, and the one on the right is rated for 24 volts, but they can both pull the plunger into the hole when the electromagnet is connected to a single 9-volt battery. At 24 volts, the larger solenoid will snap the plunger into position with great speed

Figure 2-10   *Mechanical solenoids*

and force, but since we only want to lift the plunger about a quarter inch and then drop it, a 9-volt battery will certainly do the job. I even have a solenoid rated for 120 volts AC that works fine with the 9-volt battery. Solenoids will operate on both DC and AC, so often they are simply rated in voltage, and will range in size from about the size of a marker lid, to as large as a pop can. Do not worry about the voltage and size of the solenoid, just make sure that the plunger can lift its own weight when it is placed on a table and connected to a fresh 9-volt battery. The larger the plunger, the louder the noise it will make when released, so keep this in mind if you have a few options to choose from. Your solenoid may also come with a plunger stopper or some type of linkage connected to the end of the plunger. You can remove all of this unnecessary hardware in order to allow the basic plunger to come free from the hole in the electromagnet's center.

If your plunger travels into the hole when energized by the 9-volt battery, it may fail to drop when the power is released due to friction or residual magnetism from the DC power source. This can be remedied as shown in Figure 2-11 by using a small piece of spring cut from a ballpoint pen, or even a tiny piece of sponge.

As shown in Figure 2-11, a bit of a ballpoint pen spring is cut and glued to the tip of the plunger to

Figure 2-11   *Plunger return spring*

help release it from the electromagnet once the battery is removed. You should be able to hold the solenoid upside down so that the plunger falls on to your workbench about a quarter inch out of the electromagnet hole, then pick it back up by energizing the electromagnet with a 9-volt battery. If your return spring is working, the plunger will then bang to the desk as soon as the power is removed from the electromagnet, which is the basis for our door-knocking sound. A little oil on the plunger may also help release it if it seems to stick in place randomly once pulled in by the electromagnet. When you get your solenoid working as described, bolt it into some type of enclosure so that the plunger

Figure 2-12 *Solenoid installed*

will be lifted about a quarter inch into the electromagnet, and then drop against the bottom of the enclosure.

Figure 2-12 shows my solenoid mounted inside a PVC box from a hardware store. There will be plenty of room inside the box for a small circuit board and the battery, and the thick plastic makes a good knock when the plunger bangs against it. If your solenoid is very strong using the 9-volt battery, then you may even consider adding a bolt or steel weight to the end of the plunger to further enhance the sound, but do make sure the plunger works reliably all the time. Now, let's build the two-stage timer circuit, and start messing with your friends' sanity.

Figure 2-13 might look familiar, especially if you built the previous project, since the 555 timer circuit is almost identical with the exception of the 10K value of resistor R2. With resistor R1 at 1M, R2 at 10K, and capacitor C1 at 100 μF, the timing cycle will occur about once every minute and will last only a few seconds. Because of this long timing cycle, it will be difficult to detect the exact location of the knocker, although if placed near the door, will certainly sound like an invisible person is knocking on it once in a while. After running to the door three or four times, your victim may wise up to the ruse and simply hang around to see what

is going on, but by then you will have had your fun, and pretending that you heard nothing makes it even more funny.

The timing cycle is taken from pin 3 of the 555 and fed into the base of PNP transistor Q1 in order to trigger the relays electromagnetic coil. Once the relay is engaged, it instantly charges capacitor C2, and at the same time disconnects itself from the transistor since the normally closed contacts are now open. Because the capacitor is charged first, the relay stays engaged for a few hundred milliseconds before the cycle repeats (as long as the 555 is saturating the base of Q1). The other pole of the relay is connected in reverse, so that each time the relay is engaged, power is sent into the solenoid's electromagnetic coil, and this is the magic that creates the 'knock, knock, knock' sound. Back in the days when cars had no electronics, this is how a signal light flasher worked; using a relay and capacitor wired up just like our mechanical oscillator. If you want to play around with the speed of the knocking, simply try different values for the 4700 μF capacitor, C2. Depending on how much current it takes your relay coil to energize, the 4700 μF capacitor may give you a few knocks per second, or less than one. You will have to experiment a little here, but you are aiming for a knocking repetition that sounds like someone rapping at the door. If you cannot find large value capacitors, just install several in parallel and add their values together to create a larger capacitor with a longer relay holding time. Two 4700 μF capacitors will create a 9400 μF capacitor, and this will really slow down the knocking repetition. While experimenting with the relay oscillation cycle, just remove the 1K resistor, R3 from the 555 pin, and place it to ground, rather than wait a minute for each timing cycle, and the relay will be constantly activated.

Figure 2-14 shows the completed circuit board tested and ready for installation. Notice the massive size of the 4700 μF capacitor, C2; the larger the value, the larger the capacitor will

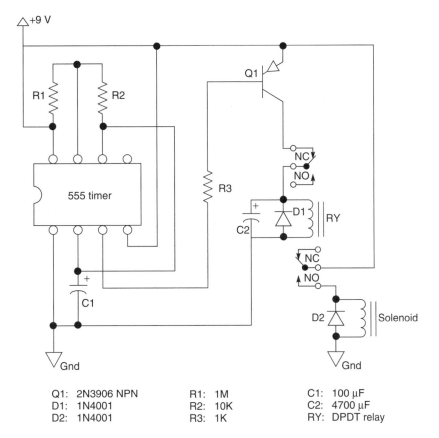

Q1: 2N3906 NPN	R1: 1M	C1: 100 μF
D1: 1N4001	R2: 10K	C2: 4700 μF
D2: 1N4001	R3: 1K	RY: DPDT relay

Figure 2-13   *Ghost door knocker schematic*

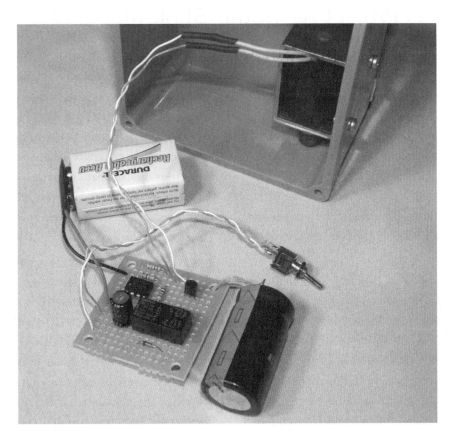

Figure 2-14   *Ghost door knocker circuit*

become, and mine is almost the same size as the battery. I also installed an on-off switch, and if you really want to make the unit adjustable, a variable resistor to alter the 555 timing cycle could also be useful. In fact, because the relay is the final switching device, practically any electrical device could be connected, from sound playback gadgets to household appliances. Now you must find the optimal placement for the device in order to make it sound like the ghost is actually knocking at your door.

Figure 2-15 shows the ghost door knocker performing its duty at the front landing of our home. The solid PVC plastic case sounds just like a knock at the door when placed against the wall near the front door, of course when you answer the door, there is nobody there, just the quiet night and that eerie feeling that the undead are restless tonight.

Figure 2-15   *Knock, knock, knock—anybody home?*

Where else could you place the knocking box that would daze and confuse your victims? The possibilities are only limited by your Evil Genius imagination.

Project 4—Putrid Stink Machine

This device is truly an evil invention that has to be personally endured in order to fully appreciate its disgusting effectiveness. I have tortured my nose for days devising an effective stink generator, and I can assure you that the project presented here can fill a room with the most putrid smell you will ever have the misfortune of inhaling. Before I get into the gross details, let's talk a little about the device and how it works.

The stink machine is another 555 timer-activated device that switches a relay on in order to carry a somewhat large amperage directly into a low-value resistor. It may seem silly to waste all that energy by feeding it directly to a 1-watt resistor, but alas, we are going to create heat from this process, exactly the same way that your stove elements create heat. This heat will then be transferred into the stink capsule for a nice slow-roasting of the stink

formula, filling a room with that very uneuphoric smell. This time you are going to need more current than can be supplied by a single 9-volt battery, so a 12-volt lantern battery, security battery or 8 AA batteries in a pack will be what you need. Figure 2-16 shows my 12-volt battery pack made from 8 AA batteries as well as the 12-volt relay, 1-watt 10-ohm resistor and a plastic bottle top.

The relay should be rated for 12 volts, but in a pinch, a 24-volt relay, or even 5 volts will probably do the job as long as your battery pack can engage the electromagnet. The resistor used to make the heat needs to be a 1-watt type with a value between 10 and 15 ohms, but you may want to experiment with other values in order to create hotter heat sources depending on your stink formula. The plastic bottle cap will be used to hold the resistor and stink formula to create what I call the "stink capsule," a ready-to-go stink-generating

Figure 2-16   *Battery pack, resistor and relay*

device that simply needs to be connected to a 12-volt power source. If you want to experiment with the best resistor value, connect the resistor of your choice directly across the 12-volt power source. However, do keep in mind that the test is destructive if successful, and your resistor will turn into a smoldering blob of carbon within 10 seconds, generating a foul smell all on its own. I found that resistors less than 1 watt burned up too quickly, and values over 15 ohms did not get hot enough to heat the stink formula properly to the smoking point. When you do find the best resistor value, it will be soldered to a pair of wires placed through the plastic cap as shown in Figure 2-17. Use a generous amount of solder in the resistor leads, or wrap the wire just in case the resistor becomes hot enough to melt the solder. Depending on your resistor value, your stink machine may work for several waves of stink release, or could burn up within a few seconds for a quick yet intense release.

Now for the fun part, brewing up the stink formula. The resistor by itself is almost potent enough to clear out a room, and it will send out a plume of smoke that smells remarkably like burning electronics—duh, it is burning electronics, buddy!

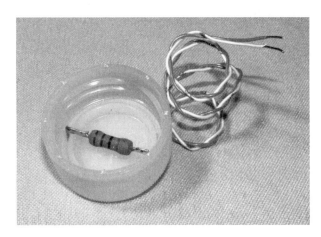

Figure 2-17   *The heating resistor installed*

Of course, I wanted to take things to the ultimate extreme, and spent days trying to figure out what would give off the most vile and rancid smell possible when the resistor began to shoulder. Plastic wrap was OK, cheese was decent, balloon rubber was even better, but still, I wanted the ultimate. Finally, I had it—hair! Yes, anyone who has played with fire and felt the fickle flames of fate can tell you that burning hair or eyebrows truly smell repulsive. Cat fur, dog hair, human hair—yes, that was the magic formula that made the stink capsule a truly mighty device indeed.

Figure 2-18 shows my perfect mix of loose dog and facial hair taken from an electric razor all nestled tightly into the plastic capsule that held the heating resistor.

Where you acquire the hair is your own business, and personally, I do not want to know,

Figure 2-18 *A truly putrid stink formula— burning hair*

but I can promise you that the burning resistor and hair combination is enough to empty a large room on the first wave, so be warned. Now, let's get the timer circuitry built so you can have the luxury of not being in the room when the device comes to life. Have a look at the schematic shown in Figure 2-19, and you will see that, once again, the venerable 555 answers the call for a simple repeating timer with a long delay between cycles. This time the two timing resistors, R1 and R2, are of equal value for a 50 percent duty cycle. This means that the relay will be on half the time, and off the other half for each timing cycle, which allows the resistor plenty of time to slow roast your stink concoction. The 220 μF capacitor C1, sets a very long timing cycle, so the machine will not come on for at least three or four minutes once the switch is engaged, making it difficult to pinpoint the source of the stink. It also allows for

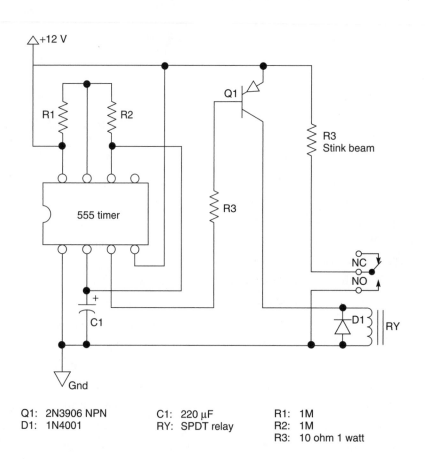

Q1: 2N3906 NPN   C1: 220 μF   R1: 1M
D1: 1N4001   RY: SPDT relay   R2: 1M
                                   R3: 10 ohm 1 watt

Figure 2-19 *The stink machine schematic*

plenty of time to get out of the room while your evil device runs its course. The relay is driven by the PNP transistor, Q1, and the heating resistor is connected to the normally open pins on the relay directly to the 12-volt power source.

If you are into self-punishment, then you could just throw the stink capsule in a box with the batteries and a switch for instant stink generation, but really, what kind of self-respecting Evil Genius would switch on a device without a timer? Again, feel free to play with the values of R1, R2 and C1 for some timing alterations, keeping in mind that larger capacitor values result in longer delay cycles. As for batteries, use only fresh alkaline or rechargeable types, since those low-cost dry cell batteries will barely have enough available current to heat the resistor. Also, before you connect the stink capsule, test the circuit with a 12-volt light bulb or buzzer instead, just to ensure that your wiring is correct. There is no point filling your own workshop with the smell of decomposing yeti, now is there?

Using my usual ultra high-quality manufacturing methods, I slapped all of the components on to a bit of perforated board as shown in Figure 2-20,

then hand wired each pin using my 10-year-old soldering iron with the blunt tip. Hey, it's not rocket science, and as long as your wiring is done correctly, everything will work perfectly, even if the end result is supposed to be burnt electronics. An on-off switch is a necessity, and you may consider labeling the thing so you don't accidentally engage it in the wrong place at the wrong time. As for an enclosure, any old plastic box with plenty of room for all of the parts will do the job. You will want to make sure that the stink capsule is vented to the outside for optimal release of fumes. The box shown in Figure 2-21 actually has a perforated lid, and during transport, I place a cork into the stink capsule to hold in all of the goodies it contains.

I don't think I have to tell you how to use the putrid stink machine—just flip the switch and get your behind out of the area, unless you plan on holding your breath for a while. As soon as that relay clicks, the smoke should begin to rise within a few seconds (Figure 2-22), and it will only take 10 seconds to fill a fair-sized room with the exquisite smell of roasting hair and burning

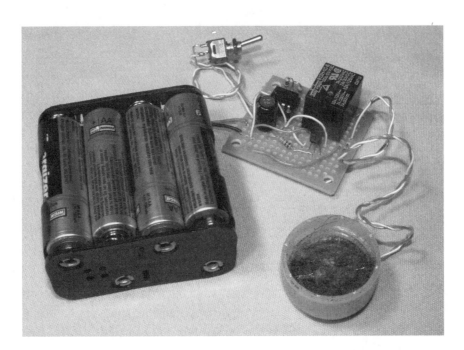

Figure 2-20   *The assembled stink machine*

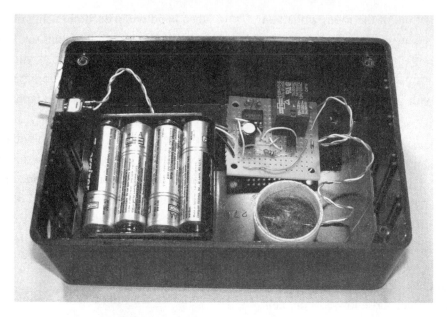

Figure 2-21   *Ready to curl your nose hairs*

Figure 2-22   *Stink machine emits a putrid odor*

electronics. By the time your unsuspecting buddy begins the painful hunt for the source of the stink, the timing cycle will probably have hit the neutral zone, but if your heating resistor holds up, you could expect several more blasts of stink to fill the room. Just remember, don't use this thing in a public area, and avoid direct placement under a fire detector, or your siege against the nose will turn into a siege against the ears instead, ruining all that hard work and spoiling the fun.

Now that you've mastered the art of creating truly annoying devices, let's continue the fun by building some annoying and truly high-tech evil critters and beasties.

Chapter 3

Critters and Beasties

Project 5—Alive and Breathing

Small critters can sometimes be vicious spiteful varmints, especially when you cross paths with them unexpectedly and they feel cornered. I'd rather take my chances with a dog over an angry cornered squirrel any day! That's why this section is devoted to furry and hairy critters that will surely scare the bejeezes out of unsuspecting humans. You could simply throw a furry mitten or small clothing item in a cupboard or a dresser drawer and wait for your unsuspecting buddy to open it and mistake the fuzz ball for some small angry vermin. But, that trick is not only old, but it will not fool them for very long. On the other hand, if that little furry beast was breathing or shaking, that would catch your victim off guard and make him or her think twice about messing with the angry little monster.

This project consists of making a blob of fur look alive by giving it a mechanical actuator that will simulate breathing or quivering. It's a very realistic illusion that appears as though the thing hiding in the dresser, in the closet or under the bed is truly alive. Practically any mechanical device that will create a little bit of quiet movement from a 9 volt battery will do the trick, and the less mechanical noise it makes the better, since the noise might give the gag away. As shown in Figure 3-1, I pulled a small magnetic actuator (solenoid) from a dead laser printer and found that the little metal plate will move nicely when the coil has 9 volts applied to it. Some solenoids have a plunger instead of a plate,

and these will also work fine, although you will have to find a way to push the plunger back out of the hole once it has been pulled in by the electromagnet.

Don't worry about the voltage rating on the solenoid, just check to see that it will move a little once you apply the power source to the electromagnet. My solenoid was actually rated at 24 volts, but since I didn't use it to move any mechanical parts, the much under-rated 9-volt battery was fine, and polarity is not a concern with these solenoids, so connect the battery to the wires in any direction you like. If the solenoid makes a snapping sound when engaged, just glue a bit of felt or foam between the metal parts to dampen the sound. If you have a plunger type that does not return once engaged, then add a spring or elastic to bring it back once the power is removed from the coil. If a solenoid seems difficult to obtain, you can probably do just fine with a toy motor by gluing some type of arm to the output shaft so that it creates a bit of lever action, although it will be fairly weak. The goal is to create a small mechanical movement that will make the fur covering look like it is pulsing or breathing, so you don't need a lot of power. The simple circuit shown in Figure 3-2 uses a 555 timer integrated circuit, two resistors and a capacitor to set a timing cycle that pulses the mechanical actuator at a rate that you can set by altering the values of the components. A larger capacitor will slow down the pulsing rate, and a smaller

Figure 3-1 *A small lever-type solenoid*

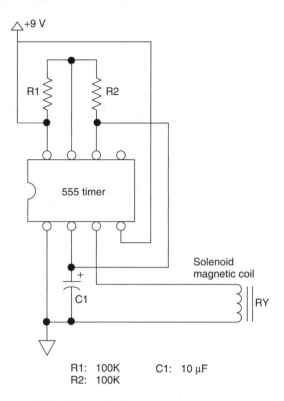

Figure 3-2 *Mechanical actuator timing circuit*

R1: 100K   C1: 10 μF
R2: 100K

value will make it a lot faster. Depending on the size of your fake furry critter, you may want a nice slow movement to simulate breathing, or a fast twitch to make it look as though the critter is shaking with fear, ready to invoke the fight or flight response.

If you need a little more response from the solenoid or motor you are using, there is no problem running the circuit up to 12 volts, but if your actuator is very low impedance, you may need to use a transistor driver circuit to source the needed current or your 555 may get hot in a hurry. Since there are only four components in the circuit, it can be built any way you like, even on a bit of cardboard by soldering the leads together after punching them through the board. I always keep a large stock of perforated board on hand, so the circuit has plenty of room for expansion on the 1.5-inch square board I have mounted it to (see Figure 3-3). Although not shown in the photo, I also added a small bit of felt to the top of the solenoid coil to stop the plate from slapping against the coil's top so that it runs silently when the battery is installed. My system starts to run as soon as the battery is installed and has no on-off switch, but if you look ahead a bit, the next few projects have light-activated circuits that could

easily be adapted to this one so that it only runs when in a lit environment such as an opened closet or cupboard door.

Mounting the mechanics will depend on the means of activation and type of critter you plan to create, which could be as simple as the small bit of fur I used or something as complex as a wire-controlled puppet-like beastie. I wanted something small and simple that could be easily placed into a cupboard or dresser with minimal set up time, so I just put the parts into a box with an open top so that the actuator arm stuck out of the top as shown in Figure 3-4. This system could be placed practically anywhere then covered with a small 1 foot square bit of fur so that the mechanical parts are hidden from view resulting an what looked like a large rat or weasel when viewed from the side. The actuator arm pumped the top of the body up and down so that it appeared the little beastie was breathing at a rate of about three times per second, giving the very

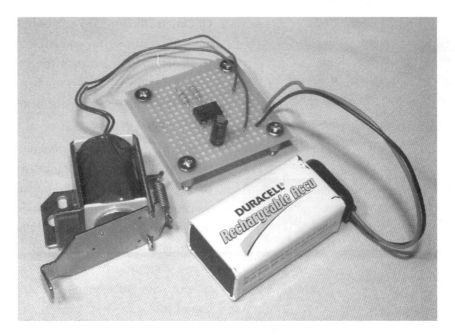

Figure 3-3 *Mechanical actuator circuit working*

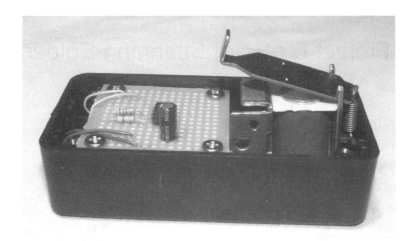

Figure 3-4 *Simple cabinet installation*

convincing illusion that it was actually alive. Most unsuspecting victims who encountered this mechanical critter weren't brave enough to hang around and inspect the furry critter, or attempt to touch it.

If you do a little creative hacking, you could make a much more realistic breathing movement by creating a tubular critter with all of the mechanics permanently mounted inside so that the actuator pulls directly on a strip glued to the underside of the fur. A good place to start might be an actual small stuffed animal with the stuffing removed so that you can build a perfectly fitting actuator to place inside. The size of the critter is also limited only to your evil imagination, and if you find a large enough battery block, you could drive a very large actuator such as a windshield wiper motor to create a moving beastie as large as a gorilla if you really wanted to give your buddies a good scare.

Figure 3-5   *What's in your cupboard?*

Of course, the tiny 555 can only source a small amount of current, so you will have to add a relay to its output in order to switch a much larger device. Figure 3-5 shows my beastie resting in the cupboard ready to startle the next person who needs a glass or mug.

This prank is very effective once your pals begin to expect regular pranks from you. At first glance, they might think that this is just you and your usual pranks, but once they see that the little fur ball is actually breathing, then they might believe that it is indeed real. Now, if you want more of an instant "in-your-face" type of scare, then keep reading. I have a full bag of tricks.

Project 6—Hairy Swinging Spider

Nothing says "hello" better than a golf-ball-sized hairy spider right in your face after you switch on a light—yikes! Grab your soldering iron and a few components from your junk box, and let's see how many of your normally "tough" friends will jump and screech like little girls when they are victim to this prank.

You could use a large plastic spider from the gag shop for this project, but the true Evil Genius is never satisfied with someone else's work, so I made my own massive hairy tarantula look alike from a bolt and some black pipe cleaners from a craft store. Thanks to the bolt, this spider was able to swing with a bit of momentum, and when it smacked you in the chest, there was no question that this was a seriously huge insect. As shown in Figure 3-6, a small ¼-inch bolt and a few black furry pipecleaners will be used to form the body and legs of the tarantula. You could also start with a ping-pong ball, or try something more elaborate

Figure 3-6   *A small bolt and some black pipe cleaners*

for a very realistic-looking spider, but that may not be necessary, since the mind will see a spider no matter how simple the prop may be once it starts to swing (as the lights are switched on), and by the time it hits the victim, he or she will probably already be shrieking with fear.

Pipecleaners are like long, thin, metal wires covered in hair-like strands, and can be found at most craft supply or dollar stores. To make the spider's legs, wrap four of the pipe cleaners around

Figure 3-7 *Wrap pipe cleaners around a bolt to form legs and body*

Figure 3-8 *This is my pet spider, Harry*

a bolt starting at the center so that there is about 2 inches at each end sticking out past the bolt to create the spider's eight legs. As shown in Figure 3-7, the overall size and shape of the legs and body look similar to a tarantula, which personally, I would not enjoy crawling on my body!

With all eight legs fastened to the bolt, take another pipecleaner and wrap it around the bolt in such a way that it creates a bit more bulk around the body. As shown in Figure 3-8, you should have a spider with a body roughly the size of a golf ball with none of the bolt showing. Again, this spider is designed for instant fright, so ultrarealism is not necessary, but if you like to go all the way, glue a few small black beads to resemble realistic looking eyes on the front of the body, and a pair of fangs

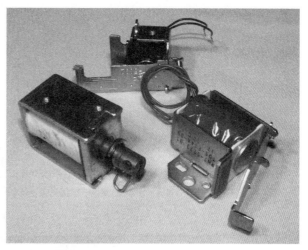

Figure 3-9 *Several mechanical solenoid candidates*

using the tips of wooden toothpicks. This type of realism might be a waste though, since the spider is triggered by the light coming on in a room, and the victim's eyes will be adjusting to the light just as the spider is attacking him or her. Not many people will be standing there trying to identify the exact species of spider after it thumps them in the chest.

As stated earlier, the spider will sit motionless in the corner of a room until the lights are switched on, at which time a mechanical actuator will release its "web" causing it to swing from its hiding place right into a doorway where the unsuspecting victim is standing. This is an electromagnetic solenoid just like the one used in the last project, or it can be a plunger type. Figure 3-9 shows several electromagnetic solenoids that would work for this project, each one able to engage from a single 9-volt battery. Since the solenoid is not doing much mechanical work, often a voltage rating of 24 volts on the coil will be fine with only 9 volts, since this will still be enough magnetism to pull the pin or plate on to the coil, although with very little force. In our design, the solenoid only has to release a very light loop of wire, so as long as the plate or plunger pulls inward at 9 volts, you will be fine with any solenoid, possible even one rated at 120 volts AC such as those found in large appliances. A dead photocopier or laser printer will have many of these small 5- to 24-volt solenoids inside among other great mechanical goodies.

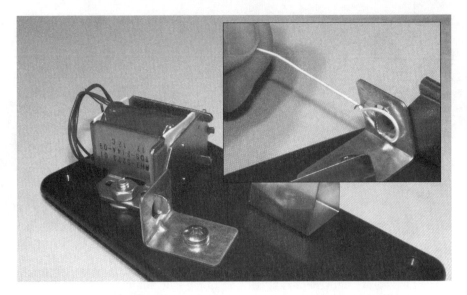

Figure 3-10 *A simple solenoid-activated release system*

Figure 3-10 shows the job this solenoid must do in order to release Harry the spider from his hiding place in the corner of a dark room. The make and style of your solenoid will determine the mechanics of your releasing device, but the simple pin-through-hole design is very easy to implement using practically any type of solenoid that pulls a pin, plunger or plate down towards its electromagnetic coil. If your solenoid has no return mechanism to push the actuator back after it has been engaged, then you may need to add a spring or elastic to hold the actuator in the up position if the friction from the little loop it is holding is not enough to keep it there. In the end, you should be able to place a loop of wire over the pin so that it is held in position until the solenoid coil is energized to pull the pin through the hole and release the loop (which will also release Harry the spider). Don't worry about the amount of noise the solenoid makes when it engages, your victim will be so busy yelping and brushing the spider off of his or her body that not much else will matter at the time.

Once you have worked out the design of your releasing mechanism using whatever parts you can find laying around your mad scientist laboratory, you will need to create the light-activated trigger circuit. This simple circuit used what is called a "cadmium sulfide photocell" (CDS cell) to monitor

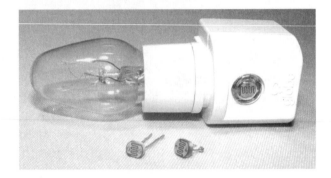

Figure 3-11 *Night lights have a CDS cell in them*

the light in a room and trigger the base of a transistor when light strikes its surface. Since CDS cells act like open switches in the dark and become resistors with an impedance of about several kilo ohms when light strikes their surface, they work perfectly as light detecting switches when fed into the base of a transistor. Most hobby shops will have CDS cells for sale, but you could always butcher an inexpensive night light for the cell. Figure 3-11 shows a typical night light and a few CDS cells that were removed from these devices. The CDS cell is easy to spot since it will have a window in front of it and look like a small disk with a zigzag pattern on it.

The light-activated trigger schematic shown in Figure 3-12 is simple, containing only four components including the solenoid and battery. The CDS cell is connected to the base of the NPN transistor so that the transistor is made to pass current to the solenoid coil as soon as the CDS cell has any amount of light striking its surface. When the room is completely dark, the CDS cell acts like an open switch and no current flows through the transistor and solenoid coil. CDS cells are not polarized, so you can hook up the two leads in any direction you like.

Just about any NPN transistor will do the job, and if you find the circuit too sensitive to ambient room light, then simply add a 100K variable resistor from the base of the transistor to ground so that you can reduce the sensitivity of the photocell. You could also place a tube made from a drinking straw painted black or some electrical tape around the photocell so that its field of view is very narrow, and then aim it

at the room light to make it more selective. With only two semiconductors, you could probably build the circuit without any circuit board at all by directly soldering the leads together, but for neatness sake, I used a small bit of perforated board and built the completed unit as shown in Figure 3-13.

Installation of the unit is very easy, and does not have to be permanent. Simply tack the launcher to the wall using a few thumbtacks, and then do the same with Harry the spider on a piece of thin thread or fishing line so that it will strike the victim once the releasing mechanism is engaged. It is easiest to first rig up the spider, and then find the best position of the launching unit by pulling back the spider so that it forms a nice smooth arc between launcher and victim. The launcher is then placed on a wall in such a way that the tiny loop of wire connected to the thread can be set so that the solenoid arm or pin will release once it is engaged. If you have your batteries connected and no power switch, then you will have to do this job in minimal lighting conditions, which is why in Figure 3-14, the battery clip is removed from one terminal of the 9-volt battery, acting as a switch. Also shown in the photo (inset) is the simple wire loop that I wrapped around the thin thread to be used over the trigger pin.

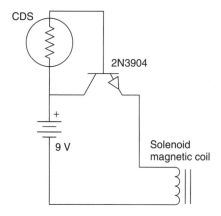

Figure 3-12 *Light-activated trigger schematic*

Figure 3-13 *Spider launcher ready for action*

Figure 3-14 *Harry the spider, ready for attack*

To test your device, load the wire loop over the solenoid release system, turn off the room lights, and then connect the battery. Once the light switch is turned on, Harry the spider should lurch into action, making an arc between his hiding spot and your body to end his journey with a nice thump on your chest. As you will have noticed, the image of the spider becomes quite clear as it travels towards you, but does not leave your mind enough time to process the image and get out of the way. If you are an unsuspecting victim at the light switch, by the time the spider is upon you, you are fully away of its hideous form and shear size, and can do nothing but scream or flee! Now you can see how tough your pals really are as they do battle with one of nature's most feared creatures.

Project 7—Carpet Crawling Creature

Imagine sitting around the living room watching TV with your pals, then without warning a cat-sized furry creature comes running across the carpet from under the TV directly at you and your friends. Of course, they all yelp and yank their feet up for cover, but you can only sit there and laugh. After all, it was you who pressed the button on the little black box that towed the creature across the floor. Yes, with this little machine, you can drag all kinds of scary vermin across the floor such as rubber snakes, furry rats, giant spiders and anything else you might think funny. This simple little hack is a tiny winch made from a toy motor and a spool of thread, and depending on the size of the motor you choose, can tow anywhere from a few ounces to a few pounds across your floor. To pull something the size of a 2-liter pop bottle across your floor, you will require a motor at least the size of a flashlight battery as shown in Figure 3-15. If the motor had a gear-reduction system connected, a much smaller motor can be used, but then you would end up with a very slow towing system, and this would not be nearly as frightening.

A good source for a medium-sized motor is fast remote-control cars, or a used inkjet printer.

Figure 3-15 *Motor and thread spool*

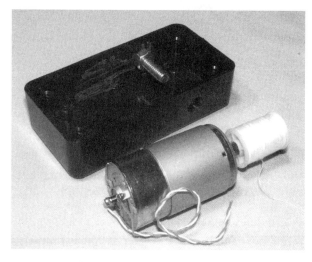

Figure 3-16 *Thread spool mounted to the motor*

Be aware of the motor voltage when looking for a suitable motor, since this unit is going to be battery powered, it would not be advisable to pull a 100-volt motor out of an appliance. The motor I used was taken from an old inkjet printer, and rated for 24 volts, although I found it to run at a good RPM and strength on a single 9-volt battery. A 12-volt battery pack would have been much better, but I wanted a small unit, and it only had to run once each time I planned to use it. You will also need a spool of thread, fishing line or wire fine enough not to be seen as it lays across the surface you plan to drag the beastie across. I like using a spool of thread because you can pick a color to match the floor or carpet, the spool is included, and it is fairly easy to "friction fit" the spool on to the motor shaft. As shown in Figure 3-16, the spool of thread can be forced over the gear that was included with the motor, and then placed in a box with a bolt to support the other end of the spool.

The bolt inside the box is smaller than the hole in the thread spool, but by only enough so that there is no friction. Since one end of the thread spool is friction fitted to the motor shaft, the other end needs to turn freely around the bolt which is used only for support. The motor will be mounted outside of the small box due to its size, so there will only be enough room in this box for a 9-volt battery and switch. If you plan to use this method,

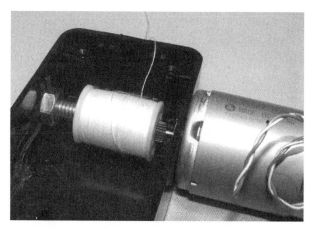

Figure 3-17 *Towing mechanics completed*

ensure that the bolt center is aligned with the motor shaft center so that there is no friction between the bolt and thread spool. Figure 3-17 shows a close-up view of the spool mounted to the motor shaft and riding on the bolt. To replace the thread spool with a different color, the bolt is removed and then the thread spool is forced away from the motor shaft. If your motor did not have a gear attached that allows the thread spool to be fitted, then dig around your parts box for some small cylindrical object that can be forced onto the motor shaft that will also hold the thread spool. In a pinch, multiple layers of heat-shrink tubing over the motor shaft would do the job, but may make a vibrating noise from misalignment as the spool winds in the thread.

The schematic for this device is probably the simplest in this book so far: just a switch, a battery and the motor. However, when connecting the wiring, be mindful of the motor's polarity so that your thread is wound into the spool, not the other way around. The thread is pulled out of the box by hand to the area where the critter will hide and then the motor will wind the thread back onto the spool when you press the button. The schematic is shown in Figure 3-18.

You don't need to place the power switch in the actual box, and it may be easier to conceal the towing mechanism if the switch is remotely placed away from the box. For a living room scenario, place the towing box under the target couch,

table or chair with the switch remotely connected via wire so that you can press it inconspicuously as your pals stare at the TV. The critter would then come blasting out from whatever furniture it was hiding under right to the place they are sitting. I wanted the switch and box as a single unit so it would be easy to implant in a hurry in any location without having to worry about wiring, so this installation works well as long as I concealed the towing box. In Figure 3-19, you can see the completed towing box with a fresh spool of white thread installed ready to drag my furry monster across a kitchen floor with light-colored tiles. You will notice that the speed of the creature will depend on the power of your motor, size of battery, and friction between the load and the floor. My cat-sized critter moves at jogging speed across a tiled or wood floor, but slows down to a walk on a carpet. Of course, the fact that this blob of fur is coming straight at you is usually enough to scare at any speed.

The bigger the critter, the better. Of course, with a 9-volt battery and toy motor, you won't be dragging a life-sized grizzly bear across your floor, so you have to do some testing to see how much weight your towing system can handle. I found that an empty 2-liter pop bottle made a perfect-sized lightweight body that would hide under various

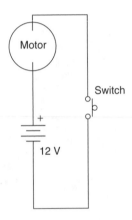

Figure 3-18   *Motor, switch and battery schematic*

Figure 3-19   *Towing system ready for action*

furniture with legs, and once covered in fur, looked like a large rat, weasel or skunk. Some furry material from a fabric store or from an old coat was glued around the pop bottle so that no part of the bottle showed, except a tiny opening where the bottle lid could be removed to fasten the thread. The thread was tied around two small holes in the bottle lid and then the lid is screwed back into the bottle. Figure 3-20 shows the initial stage of my black furry beastie using a 2-liter pop bottle, a bit of furry cloth and some glue to hold it together.

The completed 2-liter critter is large enough to make an unsuspecting person run for cover, yet light enough to get some decent speed using the 9-volt towing system. If you want to make the

Figure 3-20 *Two-liter pop bottle and fur covering*

critter a bit more evil looking for well-lit environments, then add some eyes or nasty looking fangs for a truly fear-provoking experience. Some other ideas for lightweight critter bodies might be: fur-covered balloons, a cardboard cylinder, stuffed animal with the stuffing replaced with a balloon or crumpled paper. If the victim is in a household that does not have pets, then obviously this critter is an uninvited guest.

The completed towing device and critter is shown in Figure 3-21. Sure, this beastie may not look that scary in this photo, but I guarantee that you will think otherwise if it came running out from underneath living-room furniture right at your feet when you were least expecting it! This joke can be made to last awhile if you let the beastie run all the way across a room to disappear under the couch or chair your victim is now cowering in. Everyone in the room will be standing on the furniture asking each other what the heck that thing was and where it went, so you can get in on the action and point around the room exclaiming that you thought you saw its fangs glisten as it moved again. "Dude, I think it's a rabid weasel, and it's right under your chair!" Yes indeed, this device is guaranteed to invoke the flight response in those unfortunate enough to fall victim to it.

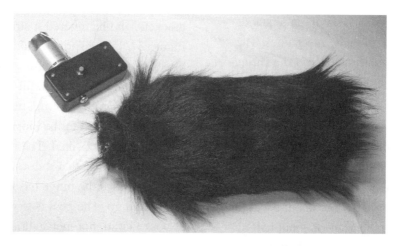

Figure 3-21 *Critter and towing system ready to scare*

This version of the furry beast prank takes things to the limit, and will freak out even your most hardened, prank-resistant friends. When this prank is in action, your victim won't have time to remember that you, the Evil Genius prankster, is likely behind it, nor will there be time to access the situation. In fact, your unsuspecting victims will only have time to duck, cover and scream with fear. Because the critter will lurch from the floor directly at the victim instantly when a light is switched on, the victim will only see the rough outline of whatever it was that is now heading directly at his or her face in a real hurry, so the effect is guaranteed to work. I call this the 'universal critter launcher' because you can set it up to hurl any small lightweight object into the air, and the mechanics can be made as large as you like. There are two parts to this system—a spring-activated hinge-launching system and an electromechanical releasing device. The two parts must be made to work together, so the amount of spring force you will be able to use will be dependent upon how much force your mechanical-releasing mechanism can deliver. Unlike the previous solenoid projects, this one will demand some mechanical force from the solenoid, so running a 24-volt solenoid from a 9-volt battery is not going to cut it. Read the entire project before you start choosing parts so you will understand what will be necessary from the mechanics. The main part of this mechanical device is the spring-activated launch mechanism formed from one or more springs and a locking gate hinge as shown in Figure 3-22.

The hinge and springs can be found at most hardware stores and available in many sizes and shapes. For a small launcher capable of hurling a rat-sized beastie from the floor to about 5 feet in the air, a spring with a diameter of about half an inch and a length of about 3 inches should do

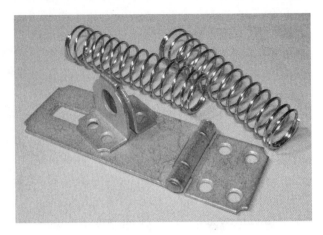

Figure 3-22 *A gate hinge and a spring*

the trick. The hinge shown in Figure 3-22 is about 4 inches in total length just to give you an idea of how large my device was. The hinge will be bolted to a wooden base so that it functions in a similar way to a swinging gate. The hinge can be locked into position by placing an object between the hole in the latching part and the hinge body. The spring will then be placed under the hinge to create a pressure that will make the hinge fly open once released. This system is shown in Figure 3-23.

Drill a hole through the wooden base for the spring, which is fastened on the underside of the board using a piece of metal with holes drilled at each end. I used one of those 90 degree shelf brackets, and hammered it straight as shown in Figure 3-24. The spring could also be bolted directly to the wooden base without drilling the hole, but this would limit the amount of compression on the spring, reducing the amount of travel delivered to the hinge when the releasing mechanism is activated. The metal plate holds the spring in place so that it does not fly away from the board once the hinge has been flipped as far as it could go. The goal is to hurl a fuzzy critter at your victim, not metal shrapnel from your launching device, so it is best to secure the spring

Figure 3-23   *The hinge is now spring loaded*

Figure 3-24   *The spring is secured to the base*

to the base. As shown in Figure 3-24, the spring is held securely in place by winding it a few turns around the metal strip bolted to the underside of the board. You may also find that the spring may need a bit of trimming if it is too long, or you will not be able to get the hinge into the locked position without bunching the spring up, which will reduce its effectiveness.

Now for the key element in this equation—the launching mechanism, which includes the solenoid, battery pack and a hinge-release pin. It is advisable to get this part of the launcher working

before attempting any circuitry, since nothing will work properly unless your solenoid has enough power to pull the pin away from the hinge once the device is loaded. As mentioned earlier, you will not get away with a 24-volt solenoid and a 9-volt battery because there is simply not enough current in that battery to energize the electromagnet, so some type of battery pack will be needed. A 12-volt solenoid with a 12-volt battery pack made of AA batteries is optimal, but you might get away with a 24-volt solenoid on the same battery pack as I did as shown in Figure 3-25. I used a bit of copper wire taken from some house wiring to create a pin that will keep the hinge in the locked position until the solenoid plunger pulls it away, releasing the hinge and the critter at high speed. It has just enough pull in the solenoid as long as the copper wire is carefully set like one would set a mousetrap, but the unit did fire reliably once I got used to setting it up. You could use a pull pin made of wire like I did, or let the solenoid pin rest directly over the hinge, depending on the type and shape of solenoid you have chosen. One thing you should keep in mind is that the pin should be set to move as little as possible and be as close to the solenoid as it can get in order to deliver the most pulling power.

Figure 3-25 *A releasing pin made of copper wire*

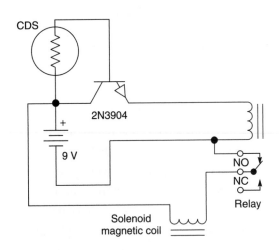

Figure 3-26 *Light-activated launcher schematic*

The farther the solenoid pin or plate has to travel, the less pull it will deliver, so a movement of less than ⅛ inch would be optimal. Test the mechanics by applying power to the solenoid to see how well the releasing pin works, then bolt down the solenoid to the wooden base once you have everything working the way it should be. At this stage, the light-activated circuit can now be added to the device to trigger the solenoid when the room lights are switched on. The schematic shown in Figure 3-26 is much like the one shown earlier in Project 6, but instead of directly driving the electromagnetic coil, the transistor first switches on a relay, which then connects the coil to the battery pack. The little transistor could not switch enough current to activate the solenoid with any real force, so it lets the relay do the job, allowing the full capacity of the battery pack to energize the coil.

If you follow the path of current through the solenoid coil, you can see that it will only activate the coil once the relay has been closed by the transistor, which will only happen if enough light strikes the surface of the CDS cell to switch the transistor on. The battery bank can range in voltage from 9 to 15 volts, but should only be enough to get the job done. The simple circuit is built on to a small piece of perforated board as usual, and then bolted to the wooden base with all the other parts of the launcher for testing as shown in Figure 3-27. The launcher has just pulled the pin to release the high-speed hinge as shown in Figure 3-27. Luckily, my fingers were out of the way because there was enough force there to leave a mark.

What you will launch from this evil device is entirely up to you, but I must admit that Harry the spider from Project 6 sure got some great air travel when placed on the hinge. You will need to experiment with weight versus distance to get things working optimally, but once you know how much mass your launcher can project, there is no limit to the evil beasties that can be flung at your

Figure 3-27 *Light-activated launcher schematic*

Figure 3-28 *Harry ready for launch, sir!*

unsuspecting victims as they enter the prank zone. You may also find that tilting the wooden base may enhance your air travel distance, but like all good ballistics experts, practice will help you reach your target. Figure 3-28 shows Harry the spider ready for launch; destination—some unsuspecting intruder's upper body. Let's see how fast they can get out of the way!

Now, don't forget to carefully load the hinge while the battery pack is disconnected, or your launcher will soon become a finger trap. The battery clip should be connected when it is dark so the device, which is sensitive to very small amounts of light, does not go off in your face. An Evil Genius should never fall victim to his or her own pranks, you know!

Here is a simple hack that can be rigged to just about any appliance, drawer, trash can, fridge, or anything with a door. This device lets you record a small audio clip and have it playback when a microswitch is triggered by the opening of the drawer to surprise someone with a silly message or sound. You have probably seen something similar to this in a novelty store in the form of a little oinking piggy that sits in your fridge, but you can build this device for much less using a sound-recording board taken from an inexpensive toy. Those totally annoying hand-held noisemakers that let kids play back a few seconds of sound is what you are after, and they can be purchased for a few bucks and easily hacked to operate from a remotely wired switch. I actually found the bare sound board at my local hobby shop for a few bucks, and you could even build one from scratch using a single sound recording IC and a few semiconductors. Just search Google for "sound-recoding IC" and you will find many different options for recording a few seconds up to many minutes of sampled sound using a single chip.

Shown in Figure 3-29 is the sound record and playback device as it came from the local hobby shop minus the addition of the larger speaker for louder playback.

To use this device, one switch is held in while you talk into the speaker to record, and the other switch is pressed to playback what was recorded. The playback switch is the switch that you will be hacking in order to attach it to a longer wire for easy hookup to a dresser or cabinet door. Since these sound devices are made as cheap as possible, there will not be much room to solder a wire, so if you cannot find suitable solder points that connect to the switch, your best bet is to remove the bubble switch and solder the wires to the contacts under the bubble. My sound board had a few resistors and capacitors, so I was able to trace the playback switch points to the legs of a few of these components to attach my trigger wires as shown in Figure 3-30. Be careful with the soldering iron as the traces are micro thin and easily damaged by excessive heating. Also shown in Figure 3-30 is the microswitch at the end of the implanted wires to

Figure 3-29 *An inexpensive 30-second sound recorder*

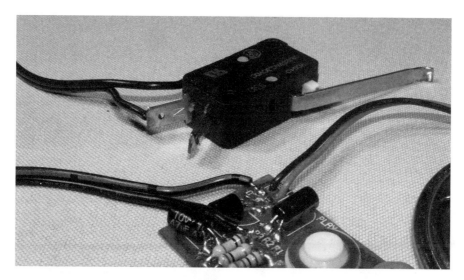

Figure 3-30   *Installing the wires onto the circuit board*

trigger the playback button. I think this switch is from the door of a washing machine, or was it a microwave oven, maybe a photocopier?

The microswitch should be easy to trigger with very little force so you can simply tape it in place wherever you want to trigger the sound. I call this unit the trash can troll because I placed it behind the garbage can so it would grumble and growl like a hungry beast each time the trash can lid was opened, but you can place it anywhere. The microswitch will probably have three connections labeled NO, NC and COM, which would indicate normally open, normally closed and common. To make a pushbutton switch, you will want to connect the normally open and common connections to your sound playback device. The guts can be mounted in any suitable container with venting for the speaker. As shown in Figure 3-31, I used an extremely high-quality plastic cabinet made from an empty container.

A little hot glue holds all the parts in place, then the container is included with the rest of the trash, but the microswitch is run along the outside of the garbage can and glued behind the lid so that opening the lid causes the switch to close and the troll to speak (Figure 3-32): "Feed me," "Don't forget to recycle," "Thanks (burp)," "What's up?"— let your imagination run wild and customize the

trash troll's comments accordingly. You'll enjoy seeing the astonished look on your unsuspecting victim every time. If the installation is temporary, use a piece of electrical tape to hold the microswitch in place, since it does not take much force to activate it.

There are many uses for an easy-to-setup sound playback device, and it could easily be mixed with any of the projects in the book to add an extra degree of realism to a prank. How about recording a shrill screech and adding sound to

Figure 3-31   *Blending in with the rest of the trash*

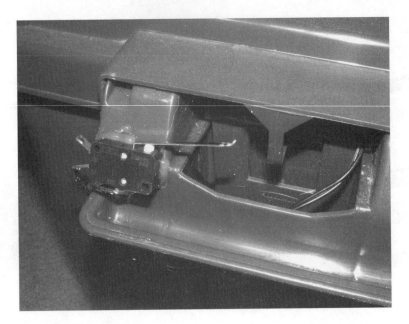

Figure 3-32 *Trash can-mounted microswitch*

Project 8 so that when the critter springs up in the air it has a spine tingling battle cry as well? You could probably directly replace the microswitch with the CDS cell to make a light-activated playback device, just like the ones that taunt a person as when he or she opens a fridge door. If you really want to scare the bejeezes out of your friends, connect the output into a loud amplifier and have it light triggered so that barrage of angry dogs start barking and howling as soon as they enter their house and switch on the lights. Maybe you could put the device inside a fake critter with a motion switch so that any movement creates an angry growl? Yes, you will know what to do with all these wonderful devices, and I'm sure it won't be hard to find that "special" person who deserves a little "Evil Genius" payback!

Let's leave the critters and beasties to rest for a bit and move on to some mechanical devices that will truly annoy and confuse your unsuspecting victims.

Chapter 4

Mechanical Mayhem

Project 10—Remote Control Jammer

Infrared remote controls make our life so much better, don't you think? When I was younger, I'd have to get up and walk across the bright green carpet, past the 8-track player and crank the huge knob to change the channel on the television. Yes, life has improved so much since those dark days, but all of that is about to change, at least for the unfortunate victim of the remote control jammer presented here. Now, before we dig right into the schematic, let me explain a little bit about how remote-controlled appliances work so we can understand how to exploit their weaknesses for our own evil needs. Most remote-controlled devices use an invisible infrared link, and they can easily be identified by the one or more infrared LEDs that stick out of the end of the remote control casing as shown in Figure 4-1.

Figure 4-1 also shows a handful of 940-nanometer wavelength infrared LEDs from my parts bin that are the same type used in most remote control units. When you press a button on the remote control, a microprocessor generates a series of binary pulses, which the infrared receiver inside the target appliance attempts to decode into one of several functions. The frequency of these ones and zeros is sent at a rate of between 38 and 40 kHz from the remote control to the receiver, depending on the make and model of the unit. I won't get into the protocol of the binary signal because that is not important here, just the fact that the base frequency is between 38 and 40 kHz. To confuse the remote control receiver, we will be sending a non-stop stream of zeros and ones at the same frequency that the remote control would normally send, but our stream will not contain any information, so the receiver will just sit there and do nothing. Because the receiver is listening to our blank pulse train, the real remote control cannot get a signal through, so this essentially blocks the remote control from working, allowing you to hijack whichever channel you like, or simply stop anyone else from using the target appliance.

The schematic for the remote control jammer is shown in Figure 4-2. The 555 timer is set up as an adjustable oscillator with a variable frequency between 30 and 50 kHz, so you can fine tune it for the most effective jamming possible. The oscillator output is fed into the base of transistor Q1, which is used to switch on and off the two or more infrared LEDs.

The reason the oscillator frequency is adjustable is because not all manufacturers use the same base frequency for their remote control protocol — why would they? In my opinion, standardization would make the world less confusing, and who would want such a thing? Anyhow, the frequency must be set exactly on your jammer, or the original remote control signal may still squeak through, since the microcontroller inside the receiver module is very good at weeding out noise and erroneous data. While you are building the circuit on your breadboard, it is a good idea to test the

Figure 4-1 *A typical remote control and infrared LEDs*

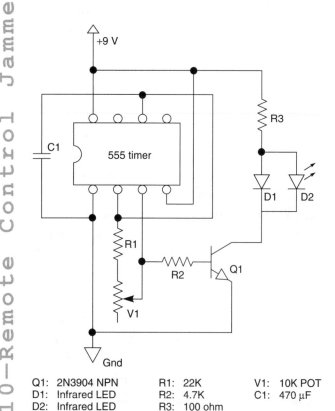

Q1: 2N3904 NPN   R1: 22K   V1: 10K POT
D1: Infrared LED   R2: 4.7K   C1: 470 µF
D2: Infrared LED   R3: 100 ohm

Figure 4-2 *Remote control jammer schematic*

frequency of the unit by measuring the output on pin 3 of the 555 timer so that you know your oscillator is working properly, and can be set from 30 kHz to over 40 kHz by altering the variable resistor V1.

Depending on the intended use of your remote control jammer, variable resistor V1 can either be a board-mounted unit or a potentiometer bolted to the cabinet for on the fly adjustments. A board-mounted variable resistor (Figure 4-3) can be set to the exact frequency using a frequency counter, or by trial and error in front of your target appliance, and then will never need any adjustments, which guarantees it will be ready to ruin your couch potato's day by simply aiming the output at the TV and hitting the switch. Of course, if you plan to use the jammer in several locations with different remote-controlled appliances, you may need the ability to tweak the frequency for best results, so the externally accessible variable resistor will be better. Remember that the frequency has to be pretty close to the target frequency expected at the receiver, and the LEDs must point towards the unit that you plan to jam, or the original remote control might still function. The ready-to-test jammer circuit is shown

Figure 4-3 *Jammer circuit ready to run*

Figure 4-4 *A couch potato's worst nightmare*

completed on a small perforated circuit board in Figure 4-3. With the jammer set to the correct frequency, and running on a fresh battery, the remote control signal will not get through, but be aware that patience must be practiced when getting it set up correctly, especially if you plan to use an external variable resistor for V1. The circuit can be built into a small black box with an on–off switch as shown in Figure 4-4 for easy concealment under furniture during stealth operations in your entertainment room.

Simply flip the on switch then place the unit under a chair or table so that the output from the LEDs is aimed directly at the target appliance, and there should be a total lock out on all attempt to control the appliance with the original remote control. The jammer will work from the same distance as the original remote, but the closer you can get to the target appliance, the better, since your enraged buddy will probably attempt to get up and point the remote closer to the set in a futile attempt to regain the power. If you want to expand on this project, you might try more LEDs and driver transistors, or higher current into the LEDs by lowering resistor R3 for extended range. Be careful when trying to pulse the LEDs with too much current or they will fry. Another nice modification would be a series of board-mounted variable resistors that could be switched on or off, so that you can hijack multiple appliances, yet not have to fiddle around trying to set the optimal frequency.

Well, I hope that you have fun watching your friends curse at their entertainment system while you sit there looking innocent. This unit does work great, but remember that the frequency needs to be set up perfectly for maximum jamming range.

Do you share a radio with a buddy or room mate whose taste in music is about as appetizing as tuna-chunk ice cream? If yes, then this project is for you, since it will convince your country-music-loving pals that their favorite radio station is on the fritz' so they might as well conform to your heavy-metal ways and move that dial over to the metal station where it should be. Maybe my musical interests are a little different than yours, but nonetheless, this project will effectively wipe out a single radio station on any FM radio that it is placed next to, allowing you to "censor" the station you least want to hear. Indeed, this device is the opposite of a radio station, since it sends an empty signal to the chosen frequency in order to wipe out the actual broadcast. Wipe out a broadcast? But, the radio station transmits at 10,000 watts, so how are we going to compete with that you ask? Well, we don't have to achieve anything close to that power level in order to wipe out the radio station; we only need to be closer to the antenna with a minimal (legal) amount of radio frequency (RF) output. If you did build a 10,000-watt radio station blocker, you could certainly block out the reception of that station for hundreds of miles, but dude, what fun is an electronics hobby when you are doing it from prison? Yes, in order to stay legal, we will only transmit at the milliwatt power level, but place our output so close to the radio that it appears to be the strongest station on that frequency. This works much the same way that an audio signal entering your ears does. The closer sound will appear to be much louder than one of a greater power at a greater distance. Because our tiny radio station is only oscillating at the desired frequency with no audio modulation, it makes the desired radio station seem like it just disappeared, much like it would during technical difficulties.

Because of the finicky nature of radio waves combined with the "dark art" of coil winding, experimenting with home-brew radio transmitters has been a road less traveled by many electronics hobbyists, even those with years of experience. Let me clear this up right now. Building a small FM transmitter is no more difficult than any basic electronic circuit, and any coils you may need can be easily made by simple trial and error using a few inches of copper wire and a bolt. I promise you that if you can make an LED blink, then you can build just about any basic FM transmitter, including most of the ones you find sprinkled amongst the millions of public-domain schematics on the Internet. Most of the small wireless microphone-type transmitters, which is what our radio station blocker is based on (AKA bugs), have only a single coil that consists of nothing more than a few turns of copper wire wrapped around a ferrite core, or a small screw. The copper wire can be taken from an old transformer or the windings on a toy motor, and you may be amazed at how tolerant the number of windings and size of the coil in your circuit will actually be. For example, if a circuit calls for "exactly 7.5 turns on a $\frac{3}{16}$ inch form using AWG #20 wire," what this really means is find an old $\frac{1}{8}$ bolt, wrap anywhere from seven to ten turns around it using whatever size copper wire you have in your junk box, then adjust it later using a ferrite screw or a small bolt. Yes, it really is that easy! I have made many small transmitters this way, and they always work as expected without having to wait weeks for some oddball coil to arrive from the Orient.

Now, let's have a look at the radio station blocker as presented in Figure 4-5 so I can get all nerdy and explain what makes it work.

If you have any understanding of basic radio electronics, then you will instantly recognize the schematic as being a simple tuned RF oscillator. If you have never looked at an RF circuit before, then let me tell you, this is a simple tuned

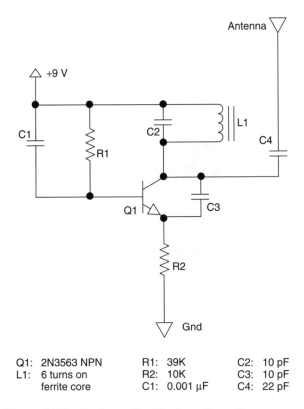

Q1: 2N3563 NPN
L1: 6 turns on
     ferrite core

R1: 39K
R2: 10K
C1: 0.001 μF

C2: 10 pF
C3: 10 pF
C4: 22 pF

Figure 4-5 *Radio station blocker schematic*

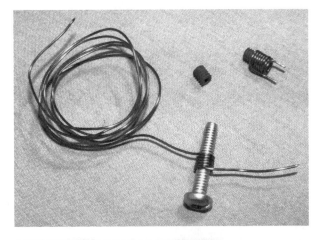

Figure 4-6 *The dark art of coil winding*

radiofrequency oscillator! Because we can adjust the ferrite bead inserted into coil L1, we have the ability to alter the transmit frequency across the entire FM radio band, which starts at 88 MHz and ends at 108 MHz. The values of the components are not that critical, and the circuit can be built on a breadboard with decent results. Transistor Q1 is any type of NPN transistor that can operate beyond 100 MHz and, if you like to read data sheets, then this value is shown as current gain bandwidth product. The 2N3563 transistor that I found in my parts box is rated at 600–1500 MHz, more than enough to oscillate at a mere 100 MHz.

I will start by showing you how to wind the coil for this project and practically any other RF circuit you may want to tackle in the future. The circuit calls for an adjustable coil with six turns, but I like to add a few extra turns when making an adjustable coil, just to play it safe. If you have too few turns, then your oscillator may run at a frequency higher than desired, but if you add too

many turns, you can always screw the ferrite bead in further to lower the frequency. In other words, a few turns too many is better than a few turns too little when making an adjustable coil. As shown in Figure 4-6, I wound six turns of whatever copper wire I had available around a ⅛ bolt. The ⅛ bolt is perfect for coil winding as it has the same threads as the ferrite bead used to tune the coil (also shown in the photo).

The small black ferrite bead commonly used to adjust a coil works by effectively reducing the number of turns in the coil as you screw it further in place. If your circuit wants a fixed coil of four turns, and you thread the bead halfway into a coil of six turns, you will probably end up with the same result, which is why this method is very easy to experiment with. Ferrite beads can be substituted by small metal screws if you are in a pinch, but ferrite beads are so much easier to adjust, and can be salvaged from just about any RF circuit board (unscrew them from metal can transformers and chokes). When you are done winding the copper wire around the bolt, unscrew the bolt, cut the copper to the correct lead length, then scrape the tips of the leads with a razor knife to remove the red or green enamel. The enamel must be removed in order to bare the conductive copper wire for soldering. Coils can be made vertically, but the horizontal method as shown in my completed circuit (see Figure 4-7) is much

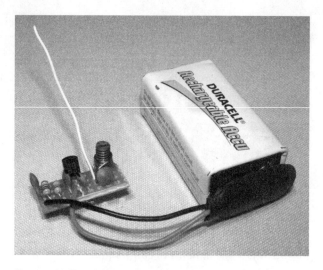

Figure 4-7 *Completed radio station blocker circuit*

Figure 4-8 *Radio station blocker ready for action*

more convenient when it comes to adjusting the small ferrite bead with a plastic screwdriver, as it will be positioned on the top of your circuit board. For coil tuning, a plastic screwdriver is much better than a metal one as the metal blade will act as part of the ferrite bead when placed close to the coil. When you think that you have the coil tuned perfectly, removing the screwdriver will shift the frequency slightly, which is highly annoying, to say the least.

The completed radio station blocker has so few parts that it can be built on a bit of perforated board less than an inch squared as shown in Figure 4-7, or by simply soldering all the components together without using a circuit board at all. If you built and tested the circuit on a breadboard, then the coil will likely need retuning, so try to tolerate your target radio station for a few more seconds while you turn the ferrite bead to the desired frequency.

The small bit of wire sticking out of the circuit board is the antenna, and it does not have to be very long due to the close proximity of the transmitter to the target radio. The longer the antenna, the further away the radio station blocker can be from the target radio—well, at least until the signal becomes much too weak to drown out the original radio station. A straight bit of wire about 3 inches long or a coiled wire approximately

12 inches long will be plenty of antenna length for this transmitter. When you have the frequency set, you can throw the guts into a small plastic case with a switch and start taking command of that radio. Figure 4-8 shows my completed unit, ready to swamp out any classical, country or pop music, and let that loud, angry, head-banging metal blast on through! Just flip the switch to the on position, and drop the box behind the target radio for hours of selective radio station censorship, compliments of the Evil Genius.

The simple radio station blocker is always ready to help you out when musical interests clash and, if you would like to take this project a little further, here are a few ideas. You could add multiple coils and switch them on and off for blocking more than one station if you have a house full of radio users with varying musical tastes. This will make your station blocker easier to use, since you will not have to open the cover to retune the coil, but it still only allows the blocking of a single station at one time. To make a multiple station blocker, you could just add more than one transmitter into a box and let them share a common antenna so you can kill multiple stations with a single device. If you want even more freedom to tune the unit, replace capacitor C2 with an adjustable capacitor with a value between 8 and 20 pF so you can tune your transmitter just like you tune a radio. You can probably find a perfect adjustable capacitor in an old transistor radio.

Well, have fun blocking radio stations and, if you decide to up the power output on this circuit, remember NOT to mention my name when the radio police take you away in their black unmarked van.

Project 12—Video Fubarizer

This little gem will drive your television-addicted buddies off the deep end, especially when you use it during special sporting events or their favorite TV series. This device works by injecting a high-frequency square wave directly into the composite video signal going into your TV from any source such as a VCR, DVD player or satellite decoder. Sure, you could simply unplug the TV, but what fun would that be? A true Evil Genius must wreak havoc in a sublime way, so as to make things look like they are simply not working properly, which is what this device will do. The interference signal can be set for light "fubarization," or all-out "fubarization," so you can tailor the level of annoyance. Now, I will leave the definition of the word "Fubar" up to you to figure out, but I will explain the details of the device at hand so you don't "snafu" the thing!

You will need a little metal can called a "clock oscillator" as shown in Figure 4-9. A clock oscillator is a crystal-controlled oscillator that outputs a 5-volt square wave at a given frequency. These oscillators only need a power and a ground connection, and will output a logic-level square wave with extreme precision, which is why they are used in many timing intensive devices such as computer mainboards, videocards, network adapters and most video appliances. You can find these oscillators with frequencies between several kilohertz up into the many hundreds of megahertz, although for our use, a frequency between 500 kHz and 6 MHz will be optimal due to the frequency of the composite video signal. Take a

look at the extremely simple schematic shown in Figure 4-10, and you will see that our device simply feeds the output of the clock oscillator directly into the video signal through a current limiting resistor and a variable resistor that we can use to set the level of interference. Since the oscillator requires 5 volts, I have included a 5-volt regulator for operation off a 9-volt battery, but if you wanted to run the unit off 4 AAA batteries and no voltage regulator instead, I would imagine that the oscillator would not complain too much about the extra volt.

The video in and out jacks are tied directly together so the device can be plugged in between the target TV and the video source. These jacks can be male or female connectors depending on your intended use. I made one of the jacks a female RCA type so that the video source could be removed from the back of the TV and inserted into

Figure 4-9 *Several clock oscillators*

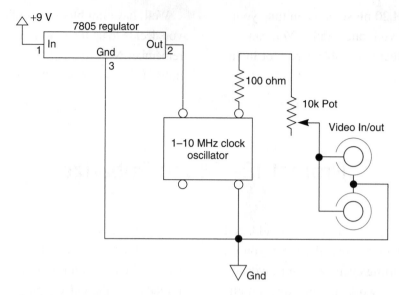

Figure 4-10   *Simple video fubarizer schematic*

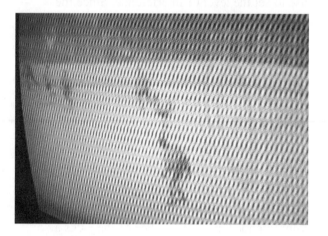

Figure 4-11   *Device working at 50 percent*

my device. The other jack is actually a bit of video wire with a male RCA connector installed so the device can then be plugged back into the tagged TV. This connector arrangement makes the unit very easy and quick to install on the target television set without needing any extra video cables. So what does this thing do to the video signal? Have a look at the fubared screen shown in Figure 4-11 and try to tell me who's winning the game!

As you can see, the video signal becomes very blurry as you inject the output from the clock oscillator directly into the line. I am using an oscillator with an output frequency of 1.8432 MHz

because this is the lowest frequency I could find in my junk box. A composite video signal contains many complex waveforms at various frequencies and, depending on the frequency of the interference signal, you will be able to generate all sorts of annoying horizontal or vertical lines in the target video system. If your frequency is too low, you may knock out the entire video signal, and if the frequency is too high, the TV may be able to simply ignore the erroneous information. Of course, experimentation is the key since we are using parts to do things they were not intended to do in order to generate signals that should never occur. The device is so simple that it can be built on to a bit of perforated board in a matter of minutes as shown in Figure 4-12, complete with the easily adjustable variable resistor and an on–off switch.

The device can remain connected to the video system when not in use since it only interferes with the signal when it is turned on and the variable resistor is turned up a bit. If you turn the fubarization knob a little bit, the lines will barely be noticeable, although a real video connoisseur may notice. At full tilt, the screen will be so messed up, it will look like grandpa's black and white TV out at the cabin, and you will be lucky to make out anything more than basic blobs and

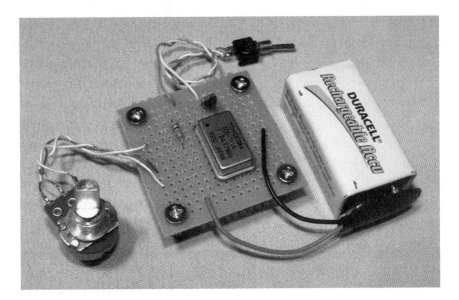

Figure 4-12 *Completed fubarizer circuit*

shapes on the screen. Now you can convince your partner that it's time to purchase a new plasma TV, or you can frustrate would-be TV watchers into doing something else while you watch the channel you wanted to watch. Yes, the uses are truly endless for such a wonderful device.

The completed device shown in Figure 4-13 can be affixed to the back of a TV in seconds by simply inserting it between the video cable and video source. The level of fubarization can then be set depending on your evil plans, be it television hijacking, or a simple Evil Genius prank. Now, if you want to take this to the absolute limit, check out the next project which can be used alone or in tandem with this one for a complete and total destruction of the TV watching experience.

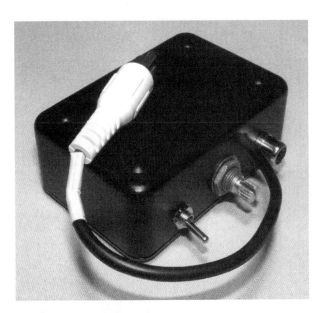

Figure 4-13 *Video will never be the same*

Project 13—Audio Distorter

This device can be used by itself or in combination with the previous project to interfere with an audio signal by distorting it to levels that would make Jimi Hendrix's amplifier sound crystal clean. Yes, I would normally install a circuit like this between

my guitar and amplifier, but this time it will be implanted between my victim's CD player and amplifier to make him or her think that the equipment or speakers are toasted. This unit works by feeding the line level audio signal through an operational amplifier (op amp) that has been set to ridiculous amounts of gain in order to distort and oversaturate the signal. A variable resistor can be adjusted to set the level of distortion from none to "brutal" by limiting the amount of dirty signal and original clean signal. Just like the previous project, the video fubarizer, this device simply drops in place by removing one of the RCA jacks that carries the audio from the source to the destination device, so it can be installed on just about any audio equipment that uses the line level RCA-style jack labeled audio in or audio output. As shown in Figure 4-14, the schematic is very simple, consisting of only an LM358 op amp and a few passive components. You could get away with just about any op amp you can find in your parts box, but the LM358 is very common, and easy to run from a single 9-volt battery, which is why I chose it.

The audio source is fed into pin 2 (inverting input) of the LM358 op amp and then taken from the output (pin 1), where it is split between the original signal and amplified signal by the 100K variable resistor. Pin 3 (non-inverting input) is tied to ground through a 10K resistor in order to set the gain of the device, which is way too high for the 1-volt audio signal fed into it. The 470 pF capacitor limits the output signal so that it is not at a voltage level that the receiving audio device cannot handle, although fully distorted. Well, that's all I can say about this device, so grab a bit of perf.

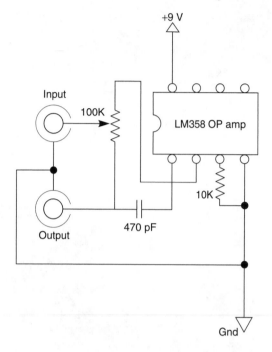

Figure 4-14 *Audio distortion schematic*

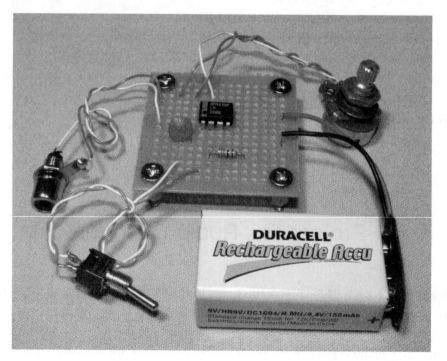

Figure 4-15 *Audio distorter ready for use*

board or simply wire all the parts together and test the unit by installing it between any line level audio source and target appliance such as a TV and VCR, or CD player and stereo amplifier. Figure 4-15 shows the completed unit ready for testing.

The results of this device set to full distortion is hard to describe, but I can tell you that nobody will be able to sit there and not notice how bad the sound is when the unit is installed. On a stereo system, the distorted audio will sound much like a blown speaker, and on a TV, the audio will sound like a very distorted copy of a movie or as if the broadcasting station was having serious technical difficulties. Over 25 years ago, you may remember the horrible audio sound on those fifth-generation Betamax movie tapes that you borrowed from your buddy—yes, that's the sound! Anyhow, in keeping with my usual standards, I finished the device using Y.A.B.B. (yet another black box) construction techniques as shown in Figure 4-16.

The case is the exact same design as the video fubarizer, since it uses the same input and output jacks, variable resistor and switch. If you want to

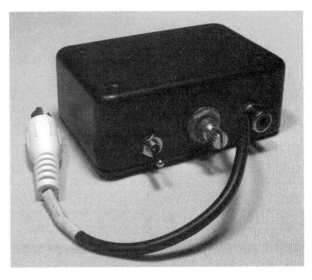

Figure 4-16 *Audio distorter completed*

interfere with both the audio and video, then both devices could be built into the same box and powered from a single battery. Once you are done driving your couch-potato pals completely around the bend, try plugging the unit between your guitar and amplifier for a truly thick and muddy distortion that would make any grunge fan ecstatic.

Project 14—Phone Static Injector

There is nothing worse than sharing a single phone line with a chatty room mate, or siblings who like to gab for hours while you are waiting for a call from one of your pals. Now, if you were a typical person with manners and such, you might just ask the offender to cut the call short so you can use the phone, but I can bet that this type of behavior is not your style! Why not use your Evil Genius skills to "hack" the chatter bugs off the phone? I can't think of a single person who would want to yap on a phone with an extremely noisy line, especially if the noise is so bad that you can't hear the other person. Enter the phone static injector!

This device injects a crackling, hissing sound into the phone line so that the incessant gabbers simply have no choice but to keep their calls short. You can control the amount of noise on the phone as well as tweak the type of noise by changing variable resistors V1 and V2. The noise is generated by exploiting the noisy characteristics of the common LM386 1-watt audio amplifier IC, and by feeding its input with its gain control pin— yikes! Hey, if it's noise you want, then it's noise you'll get. The source of the noise sounds much like a cordless phone out of range, and depending on how far you crank V1, will sound like a slight

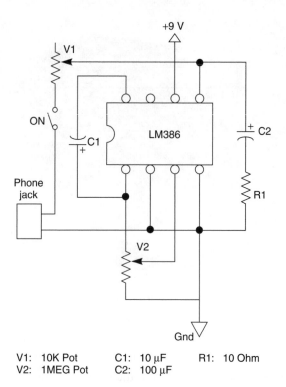

V1:   10K Pot   C1:   10 µF   R1:   10 Ohm
V2:   1MEG Pot   C2:   100 µF

Figure 4-17   *Phone static injector schematic*

Figure 4-18   *Cutting up a standard phone cord*

hiss to a total loss of signal, causing an abrupt end to the call you are hijacking. The schematic for the phone static injector is shown in Figure 4-17.

Since this device must plug into any phone jack on the wall in order to make noise on all of the phones using that line, you are going to need a standard phone cord to cut up and use as an interface cable. This phone cord can take the form of a double-ended cord to connect a phone base or modem to the wall, or it can be a phone extension cord with a male and female jack at each end. Regardless of the type of cord you butcher, it will have at least two conductors, and as many as four, although only two of them are being used. You need to cut one end off the cord so that you have a length of cable with one male telephone jack at one end and bare wires at the other. The two wires you want will be red and green, and the other two wires that are not needed will be black and yellow. Cut away the black and yellow wires, then carefully strip the insulation from the red and green pair so you can solder the tips. Whatever you do, never cut the red wire (sorry, I just had to say it)! The wire may be a funky braided material, so

soldering the end may be a bit difficult, but with a little patience it can be done. As shown in Figure 4-18, the red and green wires (center pair on my cable) are stripped of insulation and soldered so they can be used with my circuit.

The circuit is very simple, with a minimum parts count, so it can be built any way you like. I used my trusty perf. board again as shown in Figure 4-19, just before I soldered the phone jack in place. Do not work on the circuit while it is plugged into the phone jack because, when your phone rings, there will be approximately 100 volts emitted across the line, and I guarantee that you will feel a serious jolt if your fingers are holding on to both wires. To test the device, switch it on and slowly crank up variable resistor V1 to begin injecting the noise source into the phone line.

You must be listening to a phone (cordless or corded) connected to the same line somewhere in the house in order to test the unit, of course. The other variable resistor V2, can be used to tweak the noise from a nice smooth hiss, to some unreal half-digital, half-analog crackling that will surely make someone get off the phone in a hurry. Also, when you are done annoying the phone users, remember to switch off the device, or unplug it from the wall or you may not get any more incoming calls, especially if V1 is turned up enough to take the phone line to the "off the hook" state. The phone

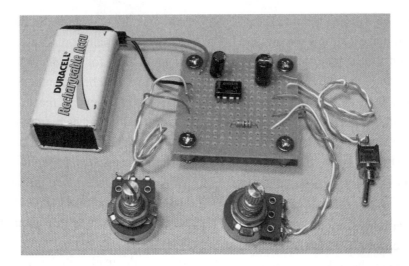

Figure 4-19   *Static injector circuit assembled*

static injector will run for a very long time off a
9-volt battery, and looks sophisticated with a few
control knobs and a black box for a cabinet as
shown in Figure 4-20.

Keep in mind that your phone company would
definitely not approve of this device and,
depending on where you live, it may be against the
law to connect such a device. Whose rules? As
with every Evil Genius device, check the laws of
the land where you live. Well, I will leave that up
to you to figure out, but as usual, don't mention
my name if the phone police manage to get past
your laser perimeter security and haul you away
for some friendly "questioning." Have fun!

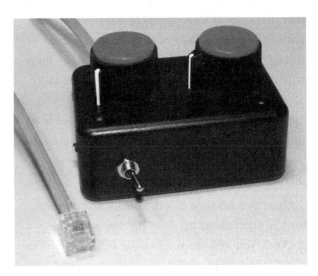

Figure 4-20   *Can you hear me now?*

Project 15—Hard Drive Failure

This practical joke is extremely effective and it
will fool even the most hard-core computer nerd,
owing to the amount of detail put into making it
sound real. If you have owned a computer long
enough, then you have likely heard the tell-tale
sound that a hard disk drive makes when it has
failed—clunk, clunk, clunk, pause, clunk, clunk,
clunk. Yes, three clunks followed by a few seconds

of silence, then three clunks again, over and over
until you realize the horrible truth—your warez
collection is history, dude! What makes this prank
work is the fact that we will be using an actual
hard disk to re-create the sound of the arm striking
the stopper as it does when the heads or disk
surface have failed. This sound is unique, and
difficult to re-create, but since a failed hard disk

drive is easy to find, we will just rip out all the hard drive's guts and make a simple circuit that whacks the arm against the stopper just like it did when it failed. Because of the extreme realism of the failure sound, this prank is perfect for the computer lab, and will always fool anyone who has heard this doom sound before, especially if they lost data due to inadequate backups. You do remember to backup your data regularly, don't you?

Anyhow, find a hard drive, any hard drive, dead or alive, and rip the cover off. This may seem easy at first but, because many hard drive manufacturers compete with each other to see who can invent the weirdest and most annoying screw head, you may need to resort to using pliers to get those oddball bolts to loosen up. As shown in Figure 4-21, I forced the top cover bolts loose by gripping them with pliers so I can finish taking them out with a screwdriver that somewhat fits into the star-shaped head. Use any means necessary to get those bolts to loosen up, including vice grips or side cutters if you have to. Oh, and yes, this certainly voids the drive's warranty.

Once you manage to get the top cover free (don't forget about the bolts hidden under the sticker), you will see the innards of a once-working hard disk drive as shown in Figure 4-22. At the top left there is a pair of magnets that create a strong magnetic field for the electromagnetic voice coil, which is placed between them. This voice coil is mounted on the end of the arm that carries the drive's read/write heads from one side of the platter(s) to the other, much like an old record player operates. Yes, I used to have vinyl records. When a drive fails, often this arm is moved back and forth several times across the entire radius of the disk platter in a futile attempt to read some data. Of course, owing to the nature of the failure (often in the read/write heads), no data are ever found, and the arm clunks off the small stopper, which prevents the heads from traveling too far off the disk platter. It is this clunking sound that we want to recreate, so the easiest way to do so will be to use the actual drive mechanics.

The only parts of the drive we will be using are the voice coil and read/write head arm, so you can keep removing those annoying bolts until that is all that remains. Remove the controller board from the underside of the drive, and then remove the motor and platters by taking out the five or six bolts that hold the plates and platter to the motor spindle (more annoying bolt heads). When you are done, there will be plenty of room left inside the drive

Figure 4-21 *Beginning the hard drive autopsy*

Figure 4-22 *Inside a typical hard disk drive*

Figure 4-23 *Only the arm mechanics remain*

casing for your battery and circuit board for a nice compact unit. Do not remove the little board that attaches the thin ribbon cable connected to the arm to the base of the drive casing, as we will need to solder a pair of wires to the end of that cable, where the controller board once sent power to the voice coil. Since the voice coil is allowed to travel freely between the two magnets, applying power to the coil will send the arm all the way to one side, causing it to bang off the stopper, making the desired noise we are after. Figure 4-23 shows what is left after all the unnecessary parts have been removed from the hard disk.

The only hard part in making this project work is finding the two points that need to be connected

in order to send power to the voice coil electromagnet. If you are lucky, the wiring will be easy to follow from the two fine wires coming directly from the voice coil, down the ribbon cable and out to the underside of the drive casing where the controller used to connect. If you have to do it the hard way like I did, then get a 9-volt battery with two bare wires coming from it and randomly test the 12 or more points in pairs until you hear the drive arm pop to one side. It may take a bit of trial and error, but you will eventually find the correct pair, and maybe if you're lucky along the way you will get to see the drive head burn up when you short it with the battery—yeah, that'd be cool. When you do find the two points that make

Figure 4-24 *Voice coil electromagnet connection points*

the voice coil work, solder a pair of wires there, and then tape them to the drive casing so that they do not break off. Figure 4-24 shows the two points I needed to connect to in order to get the voice coil to activate, and wouldn't you know it, they were the last two points I tested. Thanks Murphy of Murphy's Law, whomever you are.

Now on to the good stuff, the circuit that makes this device do its duty. As shown in Figure 4-25, the schematic for this device looks a bit complex at first, but it is relatively simple, consisting of a simple 555 timer set to oscillate at about 120 cycles per minute, and a 4017 decade counter, which allows us to set the "rhythm" of the drive clunking. A decade counter counts to ten by sending a 5-volt "logical one" to one of ten output pins for every clock pulse that is sent to its input in pin 14. If we simply connected the voice coil directly to the 555 timer's output, it would bang off of the stopper non-stop about twice every second, and that would not fool the true computer nerd who knows the sound of a failed hard disk drive. By connecting the voice coil to only the first three digits of the 4017 counter, we get three clunks, followed by seven blanks, effectively creating a rhythm from the incoming clock pulses that sounds exactly like a failed drive.

The schematic also has a 5-volt regulator to allow operation from a 9-volt battery, and is necessary owing to the 5–6-volt requirement of the 4017 counter. I tried running the works from a 9-volt source to make things simpler, but it doesn't work, trust me. The three diodes connected to each output pin stop current from feeding back into the 4017 outputs as they switch on and off in sequence. LEDs could be used in place of the standard diodes if you like, and would give you a visual feedback of which pins are active, almost like a simple light chaser. Feel free to experiment with component values to see what happens. It's all about hacking and learning by trial and error, right?

The computer hard-drive failure simulator is shown in Figure 4-26, ready to make some poor, unsuspecting, computer nerd head for the nearest backup server to make sure the most recent backup has been made. Yes, indeed, this little device sounds like the real deal when it is placed behind a computer cabinet or even inside the cabinet if you really want to go all the way. A few additions to make this prank even funnier might be a remote trigger running from your victim's desk to yours so you can be there when it goes down (literally). If you know your way around a computer, and are

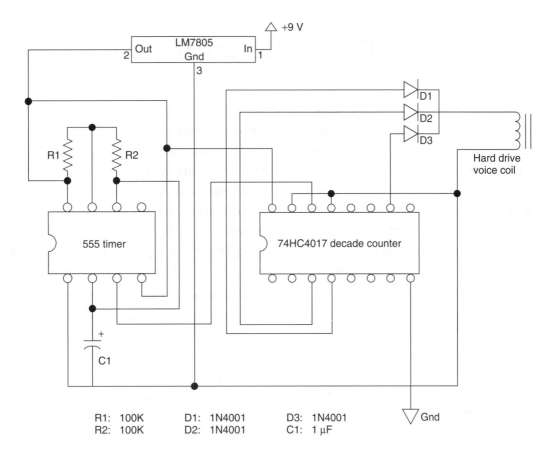

R1: 100K   D1: 1N4001   D3: 1N4001
R2: 100K   D2: 1N4001   C1: 1 μF

Figure 4-25   *Voice coil "rhythm" circuit*

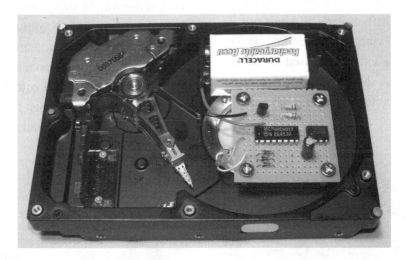

Figure 4-26   *Computer drive clunker ready to scare*

not afraid to do a little hacking, you could power the unit from the 12-volt line on one of the free power connectors inside the computer so that the clunker begins to work once the computer is switched on. If you only have a small window of opportunity to plant the device, but want to make it even more realistic, open up a full-screen command prompt and type some horrible failure message like, "Error reading cluster 956431, hard disk drive controller failure." I'm sure that you will come up with ways to horrify your computer buddies. After all, you are the Evil Genius.

This project is a simple mechanical device that makes a load banging sound much like a "monkey wrench" thrown into a finely tuned machine with many large gears. OK, that may be a little far fetched, but this thing sure can make a lot of noise that sounds like serious engine or transmission troubles requiring immediate attention. If you slip it under the seat and flip it on just as your unsuspecting buddy starts the vehicle, it will certainly make a disturbing sound as though something has just gone awry with the engine. Of course, a device like this could be used for more than just engine troubles, and it will enjoy banging away just about anywhere there is a machine with moving parts that normally does not sound like a bucket of bolts caught in a blender. For this simple mechanical project, you will need a DC motor of some type and some way of connecting a weight on the end of the motor shaft. The weight will throw the motor off balance since most of the mass is on one side of the shaft, causing all kinds of beautiful clanking and vibrating, depending on the nature of the mounting and cabinet used to contain this evil work of art. As shown in Figure 4-27, a common small DC motor will do the job and,

the larger the motor, the louder the sound will be. I have chosen a common 3–6-volt toy motor that I will be running at 9 volts to get some serious bang out of my device. A larger motor such as the type found in remote controlled (RC) race cars will make a serious amount of noise, but will also require a much larger battery than a simple 9-volt type. The choice is yours, and will most likely depend on what's in RC your scrap parts pile.

Also shown in Figure 4-27 is a bolt and the inside of a large twist connector. This will be my method of adding the out-of-balance weight to the end of my motor shaft, and again, this is just a hack, so use whatever means you like to get some type of out-of-balance weight on to the end of that motor shaft. The size of the weight will depend on the power that the motor can deliver and the type of cabinet used, so don't try to swing a bowling ball around with a 200-milliamp toy car motor. Also, if the weight sticks out too far off the motor shaft, it may not fit into your enclosure, so take this into account when you are testing the parts. My bolt and fastener, shown in Figure 4-28, is about the largest weight I could get away with on this size of motor and, when powered by a 9-volt battery, the bolt spins around and vibrates the

Figure 4-27 *A small DC motor and weight*

Figure 4-28 *Mounting the weight to the motor shaft*

motor so much that it's almost impossible to hold on to.

Once you have some type of weight attached to the motor shaft to create a good vibration, fasten the motor to some type of cabinet to contain the unit. There are two ways to do this, but each will give a different result and type of clunking. The simple method is to fasten the motor to the inside of whatever cabinet you plan to use so that the weight is allowed to spin freely while the motor is held solidly in place. This method will not make much noise, but will vibrate the cabinet and whatever it may come in contact with so that the noise is generated by the actual shaking of your target machine. The second method, which is the one that I used, makes a lot more noise by letting the motor flail all over the place while randomly striking the sides of the cabinet with the spinning weight. This method makes a truly nasty sound that will make your victim's eyes light up when he or she hears it. This noisy method requires the motor to be mounted on some type of spring or rubber pad so that it can jostle around a bit, letting the weight strike the insides of the cabinet randomly. Figure 4-29 shows my spring-mounted motor mount made from a pair of bolts, a spring and a wire wrap. Again, use some creativity, and whatever parts you have collected to make your system work.

The rest of this project is easy. Just solder some wires to your motor, a switch and the battery so you can turn the horrible noise-maker off once the joke is over. Figure 4-30 shows how I mounted the parts on to the lid of my cabinet so that the guts can be dropped into the box when the lid is installed. If you choose the spring-mounted motor system, ensure that your wires are securely soldered to the motor terminals, and try to use a wire that can take the repetitive flexing it must endure as the motor vibrates all over the place. Some careful tweaking of the motor mounting position in the cabinet may also be necessary to avoid having the weight come too close to the

Figure 4-29   *A spring mounted motor for serious noise*

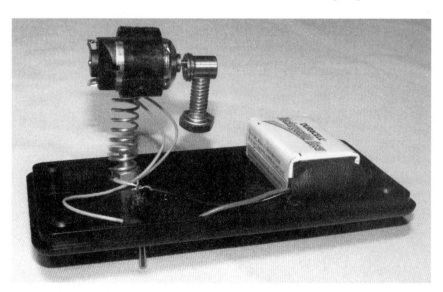

Figure 4-30   *Parts ready for installation*

inside walls of the cabinet, or it may simply get stuck. What you want is some good striking action, but also some good motor revolutions to get the weight up to speed.

The cabinet is basically empty, but that little motor bounces around so much that I needed all of that room to keep the weight from becoming stuck inside the cabinet. Figure 4-31 shows my final product, ready to make anyone's car, motorcycle, photocopier or even school locker sound like it has become possessed by 1,000 angry demons. Yes, this device can really throw a monkey wrench into the gears, or so it would sound.

I hope that you enjoyed the mechanical mayhem chapter. I know I had a great time building and testing out all these wonderful devices on my unbelievably patient friends. Feel free to mix and match some of the technologies and ideas presented in these and other chapters to create your own truly evil arsenal of high-tech practical jokes.

Figure 4-31 *The noise this thing makes is exquisite*

After all, getting even with your buddies for previously played pranks is war, my friend, and you must fight! But, remember, it's all in good fun.

In the next chapter, we will create more mayhem with various devices and things that go bump in the night.

Chapter 5

Things That Go Bump in the Night

Project 17—Glowing Blinking Eyes

Have you ever seen those funny scenes in cartoons where a person is trapped in the dark, then all of a sudden glowing, blinking eyes pop out of nowhere? Well, that's what this next project is all about, but it works in the real world. This prank works best in a large dark room, or at night time outside in a forested area because the eyes will look very realistic at a distance, especially when they start to blink. The key to this project is the pattern that an LED will make on a diffused surface such as a ping-pong ball that looks like an iris from some nocturnal animal reflecting the moonlight. The added realism of the blinking shutters will make most people turn around and run if they spot the eyes from across a dark room or peering at them from the dark surrounding forest. Strap the eyes to a tree behind the outhouse, or place them at the end of a long dark hallway, and watch your friends scatter in fear!

For this project, you will need two white ping-pong balls and two bright LEDs, preferably of the high-intensity white variety. Any color ping-pong ball and LED will indeed work, but only the white LEDs and surface will look like real eyes in the night, unless your friends believe in red-eyed demons. Figure 5-1 shows the parts used for this project—the two bright white LEDs and a pair of white ping-pong balls.

The ping-pong balls will be cut in half to be used as diffusion screens for the LEDs, but since most of them have a logo on one half, you will have to dice up a pair of them to get two blank halves. The ping-pong balls are very thin and rip easily, so start by digging a small hole in the side along the equator using a drill bit turned by hand so you can have a place to start cutting. Place the tip of a very sharp small scissor blade into the hole and try to cut along the visible equator, dividing the ball into two equal parts. You will probably have to trim the ball down a little more to get the best LED pattern on the inside, but it's best to start with half a ball and remove a bit more material later. Take one of your LEDs and power it up with a small 3-volt button cell so you can see how far away from the inside of the ball the LED should be to project the best-looking "iris" on to the surface. As shown in Figure 5-2, the beam of light from the LED will project a pupil and iris on to the diffused surface of a ping-pong ball when held at an inch or so away from the top of the ball. Don't worry about overdriving the voltage on the LED with a button cell, as there will not be enough current available in the battery to hurt the LED, and it will probably not be nearly as bright as it would be with proper power supply and less voltage.

Next, find a suitable plastic box that can hold the two LEDs and ping-pong ball halves at a distance of between 5 and 6 inches apart, as well as having room for a small circuit board and 9-volt battery. The LEDs will be mounted to the cabinet through holes drilled in the top and then the two ping-pong ball halves will be glued over the LEDs.

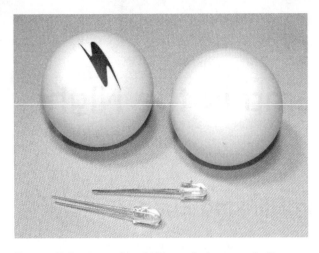

Figure 5-1   *Two white LEDs and ping-pong balls*

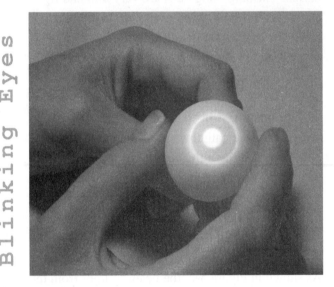

Figure 5-2   *Projecting an iris and a pupil on to the ball*

Some experimentation will be needed in order to find the perfect depth to cut the ball halves in order to project the best-looking iris once the LEDs are mounted in the cabinet, which is why it's best to start with exactly half a ball and then trim a bit off until you have something that looks good. I found that the ping-pong balls needed about ¼ inch off past the equator for the LEDs I am using, but keep in mind that the field of view on an LED can vary quite a few degrees, so test this for yourself. Usually, the brighter LEDs (rated in millicandela (thousands of candela)) will project a sharper iris, but require less distance from the surface of the ball, owing to the narrow viewing angle. As shown in Figure 5-3, the LEDs are mounted through holes in the plastic box, and the position of the ping-pong balls over the LEDs has been traced using a black marker.

Now let's create the mechanical eyelids to make the evil eyes look much more convincing. Sure, you could just strap the glowing eyes to a tree like they are, but they may only fool the most paranoid of victims, so the addition of a little mechanical motion will be sure to get them all. The eyelids are nothing more than dark shutters that are repulled over the glowing eyes at a rate controlled by a timer circuit, but at a distance in the dark they do indeed look like blinking eyes. The shutters are made by forming what looks like the frames of a pair of eyeglasses using a bit of copper wire taken from some house wiring. The shutters connect to

Figure 5-3   *Mounting the LEDs in the plastic box*

the box at each side by sticking the end of the wire into a pair of drilled holes so the can freely move up and down over the ball halves. The eyelid parts of the shutters are made by soldering two small loops of the same copper wire on to the main wire so it ends up looking something like the one I made in Figure 5-4. Feel free to make the eyelids any shape you like to convey a little emotion. Angled upwards, the eyelids make the eyes look

evil and angry, just like those cartoon eyes I was talking about earlier.

The actual eyelids are made from black electrical tape cut around the shape of the eyelid wires as shown in Figure 5-5. Actually, any dark material that you can tape or glue to the eyelids will work, as long as it blocks out all of the light when placed in front of the ball halves. Also shown in the photo is an elastic band tied around

Figure 5-4 *Making the eyelid shutters from some copper wire*

Figure 5-5 *Eyelid covers and elastic band installed*

Figure 5-6 *Connecting the motor arm to the eyelids*

the center of the shutter so that they can be pulled up and down by the motor actuator that will be installed shortly. Cut a small elastic band in half, then tie one end to the center of the shutter.

The up and down movement of the eyelid shutter will be delivered by a high-tech rotary actuator with a right-angle drive system—this is also known as a "toy motor" with a bit of copper wire soldered to the shaft. You can use any motor with enough power to lift the shutters up and down using the elastic band as a towing wire, so look for a small electric motor from a kid's toy. As shown in Figure 5-6, a small actuator arm is made by wrapping a bit of copper wire around the motor shaft then soldering it in place so it does not slip. The elastic is then tied to the end of the actuator arm so that, at full rotation, the eyelids uncover at least 80 percent of the ball halves, and when no power is applied to the motor, the shutters drop back over the eyes. A little experimentation with the length of the copper wire arm and the elastic will be necessary, so make things longer than they need to be at first and trim them down as needed.

Also shown in Figure 5-6 is a small wood screw placed at one side of the box to restrict how far down the shutters will travel when there is no

power applied to the motor to lift the actuator arm. This wood screw stops the eyelids from dropping below the eyes to create a more natural blinking illusion. If you have trouble tying the elastic to either part of the wire, just do your best, then use a bit of hot glue to secure it in place. Now, your motor should be able to lift the eyelids up and down at levels that look like realistic blinking from a distance in the dark. Avoid dropping the 9-volt battery directly across the motor at this point, or you may bend or loosen the copper arm from the torque. It's best to try the motor on less voltage or through a current limiting resistor of 10 or 20 ohms. The schematic for the eye blinker is shown in Figure 5-7, and it uses a simple 555 timer circuit to drive the motor for about half a second and then repeats the cycle about once a second. This timing cycle can be altered by changing the values of resistors R1 and R2 or by increasing the value of C1 for much longer blink cycles.

Resistors R3 and R4 limit the current to both the LEDs and the motor, and should be chosen based on the components you plan to use. If you do not know the current capabilities of the LEDs you have, then start with a higher value for R3 such 1K, and work your way down to increase the

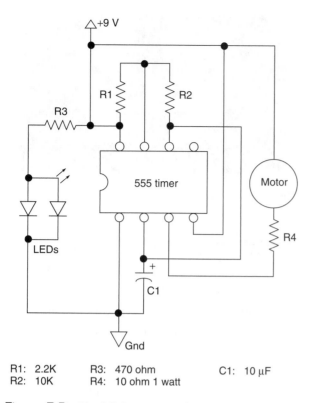

R1: 2.2K   R3: 470 ohm   C1: 10 μF
R2: 10K   R4: 10 ohm 1 watt

Figure 5-7 *Eye blinker timer schematic*

brightness of the LEDs if necessary. Keep in mind that brighter is not always better or your eyes will not look like they are reflecting moonlight—they will look like obvious fakes. The eyes should be visible at 20 feet away, but not shine so bright that they look like yard lights. In my design, I drove the LEDs at about 1/10 the brightness level they were capable of if full current were applied. R4 also limits the current that will be delivered to the motor, and since the 555 is directly driving the motor, this is necessary. Normally a 555 timer would not be used to drive a motor, owing to limited current sinking capabilities, but we only want enough torque to lift the shutters so the copper arm does not slam in the box cabinet when it reaches full rotation. Also, the motor will be in the stall position for a small amount of time, but that also creates a large load on the timer, another reason why R4 is very important. If you find that the timer warms up after a few blink cycles, or the eyelids pop open with way too much force, then increase the value of R4 until things are working properly. The simple circuit is assembled on a bit of perforated board as shown in Figure 5-8, ready for installation into the little black box.

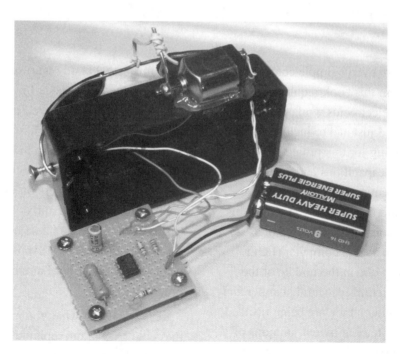

Figure 5-8 *Circuit board ready for installation*

A decent-sized magnet glued to the back of the cabinet will make mounting the object on metal surfaces very convenient, as well as a pair of hooks at each side of the case for wrapping it around a pole or tree. The best height for the device is about 4 feet off of the ground to simulate the size of a wolf or some other large nocturnal creature that can hold its ground against a full-grown human. Another little trick you could use to make the illusion even more convincing for outdoor use is the installation of the device on a spring-loaded pole so that it wavers a bit in the wind, making it look as though the creature was trying to get a distance bearing on its prey (you) before attacking. Figure 5-9 is a shot of my blinking eyes as they peer back at me deep in the dark corner of the storage room. Would you hang around to investigate?

Besides the faint clinking noise that the eyes make when they blink, the illusion is pretty eerie to look at, and at a distance, there would be no audible noise at all. If you find that the motor is a

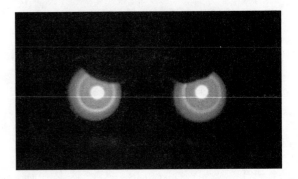

Figure 5-9 *Are you looking at me?*

bit loud, then you could try increasing the value of R4 to limit the motor current a bit more, or try adding a bit of sponge at the points where there is mechanical contact to lessen the click. At 20 feet or so, the unit does not make any noticeable noise, and this seems to be a perfect distance to make the eyes look real. So next time you are out at camp telling stories of blood-thirsty wolves and man-eating bears, add these glowing, blinking eyes into the mix for a real good scare. The first person brave enough to use the outhouse at midnight will be back in a real hurry, I bet!

Project 18—Computer Audio Nightmare

This project may seem overly simple compared to some of the projects in this book requiring hardware hacking or electronics work, but I find it to be so effective that I just had to include it. You will only need one thing for this project, a computer with sound capabilities and any microphone, and the entire prank is done using tools already built into the operating system. This project is called the computer audio nightmare because it uses your personal computer (PC) to playback a loud audio clip in the middle of the night, running your victim's peaceful slumber and making it sound as though they are being invaded by a SWAT team, or attacked by evil minions of the night. I came up with this project to have a

little fun with a buddy who was watching our house for the night, since I owed him one for some other prank that was played on me in the past. I recorded what sounded like a crazed lunatic smashing through the window and then added some pitch change effects to make it sound as evil as possible. The speakers on my computer were cranked up very loud so that, when the sound was triggered at 3 am, the "fake home invasion" would sound like the real deal. Apparently, the prank went over so well that I am expecting payback any time now!

Let's start with what you need to record sound into your PC—a microphone with a jack compatible with the input on your sound card.

As shown in Figure 5-10, most PCs will have a ⅛ stereo plug for a microphone or input device, and will often have a photo of a microphone or an arrow pointing towards the jack indicating that this is the plug for your microphone. I don't know if there are any color standards, but the microphone jacks on the PCs in my house have a pink outline.

Some PCs and many laptops may have a microphone built right into the case somewhere,

so there may not be an external connector, but that is OK, as long as there is some way to record sound into the computer. If you are familiar with multimedia software and sound recording, then you are probably already thinking about ways to create scary sound effects for this gag, so feel free to browse ahead now that you understand what this project is about. If you have never tried to get sound into your computer before, then find a microphone that will fit into the microphone or input jack on your computers sound card. Often, these simple microphones are included with the computer, and can be purchased at any computer store for under $10 (they are called multimedia microphones). Once you have the microphone plugged into the input jack on the computer's sound card, open up the volume control by double clicking on the little speaker icon at the bottom right of your task bar. These instructions are based on Windows XP, since it is the most widely used operating system at the time this book was written, but you will be able to follow along with just about any operating system (Windows Vista also has the same programs and controls we need). Figure 5-11 shows the panel that will appear once you double

Figure 5-10   *A microphone input jack*

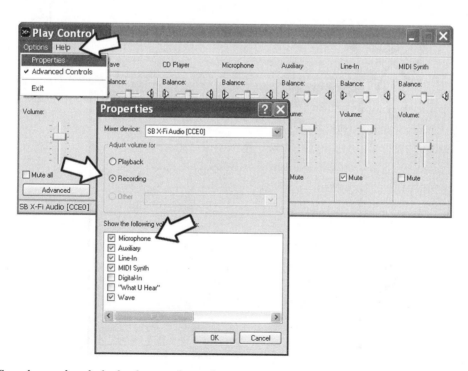

Figure 5-11   *Sound record and playback control panel*

click on the volume icon, or engage it from the start menu.

As shown in Figure 5-11, start by clicking on "Options," then click on "Properties," and finally click on "Recording" so you can make sure that the button "Microphone" is checked off, indicating that the microphone input is the currently active sound input port on the computer's sound card. Once "Microphone" is selected, press OK to bring up the microphone input properties panel as shown in Figure 5-12.

The control panel shown in Figure 5-12 lets you adjust the input levels for your microphone, and you will want to ensure that everything is "cranked" to the limit in order to get a loud enough recording. Start by moving the microphone slide all the way up and then click on "Advanced", if your panel has such a button, to reveal the microphone boost panel, if available. Again, crank all sliders to the max, and check off "Microphone Boost", if such a button is available on your system. Now you can click "OK," or apply through all of the panels in order set and close your microphone properties. Your microphone is set to be the default sound input device, and its volume level will be set to the maximum level so you do not have to scream at the top of your lungs in order to make a decent recording, although you may want to anyhow! Now, let's start up the basic sound recorder that is built into your operating system so you can test your microphone and make

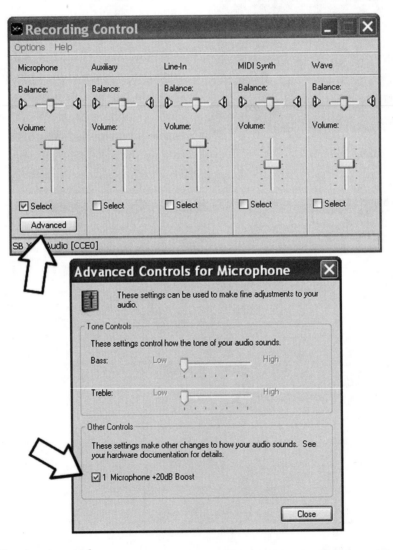

Figure 5-12 *Recording input controls*

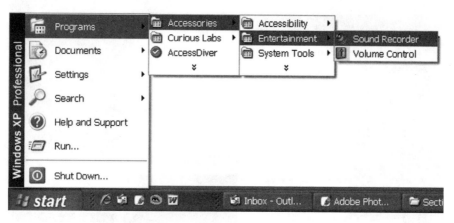

Figure 5-13   *Running the basic operating system sound record program*

basic recordings. As shown in Figure 5-13, click on "Start," then "Programs," then "Accessories," then "Entertainment," and finally on the program called "Sound Recorder."

This sound recording program is as basic as it gets, but will be more than enough to record whatever horrific yelps, screams, or barks you want to use to wake your slumbering pals up in a hurry. You can even add a bit of echo, or slow the sound down to make it sound like it came spewing from an evil demon by lowering the pitch of the recording. The sound recorder program will open a tiny window with a few options such as record, effects and file functions. Go ahead and click on the record button as shown in Figure 5-14 to start recording sound from your microphone.

Any sound heard by the microphone will be translated into a wave shape in the little sound recording window as shown in Figure 5-14, as I cackle and grumble into my microphone, reciting my favorite lines from the *Evil Dead* trilogies for later playback. If your microphone is properly connected and your volume control set to monitor the correct input, then you should see a nice saturated wave file as you speak into the microphone at a moderate level. If something isn't connected correctly, then your PC's hard disk drive is probably laying on the other side of your room in a large plume of smoke as the rest of the computer incinerates in a huge red fireball. No seriously, unless you are in the wrong input

Figure 5-14   *Recorded sound is shown as a wave*

plug, everything should be working fine, but if there is no modulation in the sound recorders window, simply try the other input lugs until you see something. If that does not help, re-check the record input control to ensure that the microphone input is selected and cranked up. Practice making the most terrible sounds you possibly can, maybe throwing in a bit of echo, or a pitch drop to mess things up to the max, and then save one of the sounds using the "Save" commands on the menu bar. Save the file to a place on your hard disk that is easy to find, since we are now going to run the scheduling tool to trigger it. As shown in Figure 5-15, the path to the built-in "task scheduler" is a long one, but I assure you that every version of Windows has one of these applications, and it will let you run any program, sound or video file that your mouse can click on, and it will do it at any time you tell it to.

On XP, the program is called "Scheduled Tasks," and will bring up a new window like the one

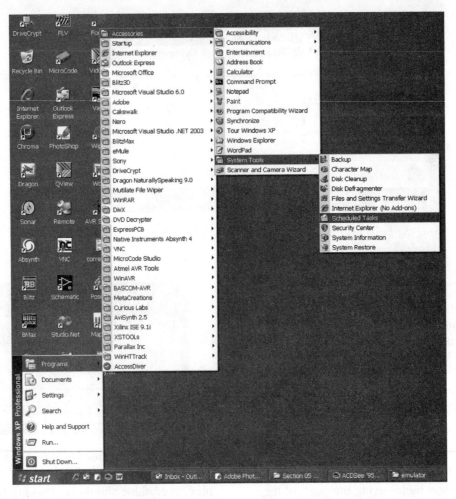

Figure 5-15 *Starting the scheduler program*

shown in Figure 5-16. All you have to do in order to schedule the sound to play at some specific time is to click on "Add Scheduled Task," then "Browse," and then locate the sound file you have just saved, and click on it. In Figure 5-16, I have selected the file "FBI-Raid.wav" as the sound file that will be blasted through the house at wee hours of the night on volume levels that will make it sound as though it were really happening.

Normally, the task scheduler would be used to trigger an application, but you can choose a sound file due to the fact that your operating system will spawn whatever program is associated with the sound file to play it back. Basically, whatever happens when you click on any file with your mouse pointer is what will happen if you choose it as the target for the task

scheduler program. The only step left is to set the time you want the event to happen as shown in Figure 5-17.

You can set the event to occur only once (recommended), or at any given interval if you like, and then the actual time is entered in the last box as shown in Figure 5-17. Once you select a time, press "Next," and the task will be set to run as long as your PC is powered up when the chosen time comes around. You might want to choose a time such as 1 minute from now just to test your volume and make sure the sound is going to trigger properly, as you can always reschedule the event again by right clicking on it and entering another time. Now your PC is ready to scare the wits out of your buddy in the middle of the night—a true computer audio nightmare indeed!

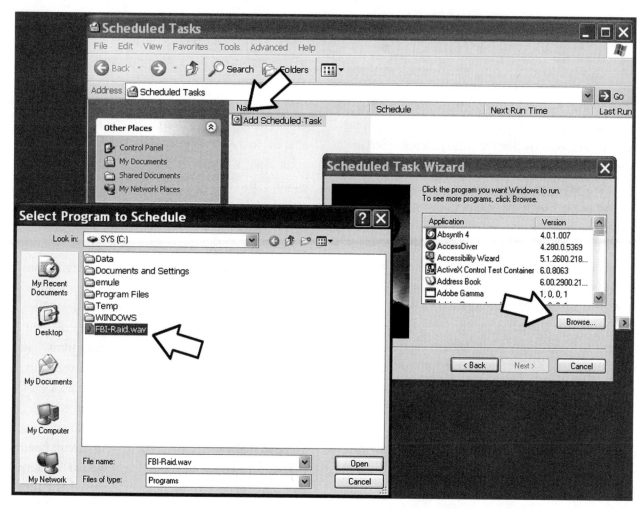

Figure 5-16   *Scheduling a sound file for later playback*

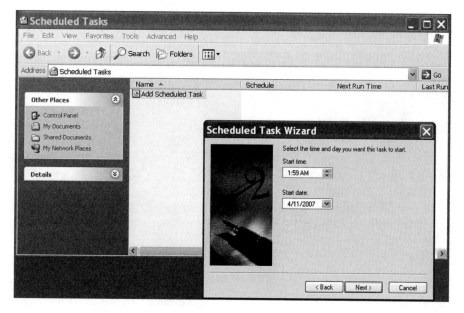

Figure 5-17   *Choosing an event time*

If you have ever had a furry little critter invade your home, then you are probably familiar with the aggravating sounds that the little rodents make in your walls at night as they scurry about. Well, this little device will closely simulate the same sound a mouse or rat would make in a wall, and will only operate at night, so it's very difficult to track down. Even the light of a flashlight will make the device go silent, so your unsuspecting victim's rodent hunting expedition will likely fail. This project uses a 555 timer set up as an oscillator that drives its output into a small open-core electromagnet in order to vibrate a tiny permanent magnet floating inside the coil. The result is a random sound resembling the sounds that a small animal would make as it gnaws at wood or some other hard material. For this project, you will need some thin, enameled, copper wire to wind an open core coil. This copper wire can be found in any small transformer, toy motor, solenoid, relay and many other small electromagnet devices. You can also purchase spools of this wire at many hobby shops for a few dollars. The exact size of the wire is not all that important. Figure 5-18 shows some copper

wire I salvaged from an old solenoid and a tiny square magnet taken from a magnetic name plate of some sort. The magnet only needs to be the size of an aspirin, or even a fragment of another larger magnet since its only job is to rattle around a bit.

To wind the coil, find a cylindrical object slightly larger than the magnet with a smooth surface so you can wind the coil and then slide if off easily when completed. For my magnet, an AA battery was the perfect size, and would allow the wound coil to be slid off without any trouble. You will want at least 150 turns in your coil, and if you have a lot of wire, use 200 or more turns. This may seem like a lot of wire, but 150 turns is only about 20 feet of wire, and it takes less than a minute to make the entire coil. Wind the wire tightly, and leave at least 6 inches at both ends so you can make the connection to your circuit board. Figure 5-19 shows the completely wound 150 turn coil ready to be slid from the battery used as a form.

Be careful when you remove the coil from the form, as it may tend to unravel once the form is no longer holding it tightly together. As shown in Figure 5-20, a few turns around each side of the coil using some of the leftover copper wire will hold the unit together so you can handle it without it unwrapping. Now you can test your coil and magnet by placing the magnet into the center of

Figure 5-18 *A tiny magnet and some spare copper wire*

Figure 5-19 *Only 150 turns on a battery makes the coil*

Project 19—Rats in the Walls

Figure 5-20   *Finished coil and a tiny NIB magnet*

the coil and applying power to the leads with a 9-volt battery. The magnet should jump around a bit as you apply and then release it. Do not hold the battery on the coil too long, or you may find it getting very hot due to its low resistance. If you find that there is not enough magnetism to move the magnet even a little bit, then you may need to make a coil with more turns, or try a much stronger magnet. A black ceramic magnet is many times weaker than an equal size NIB (neodymium) magnet, so you may want to try one of those instead—they are usually silver in color.

Once you are satisfied with the amount of movement from your coil and magnet, build the 555 timer circuit shown in the schematic (Figure 5-21). This 555 timer will pulse the coil at a rate determined by C1, R1 and R2, so feel free to experiment with the values to get the most realistic sounding critter you can. The 555 timer will only start if there is no light hitting the surface of the CDS cell, or while it is removed from the circuit altogether for testing purposes. Ensure that the 555 timer does not warm up while the unit is running or you will have to either add more turns or a resistor in series with the coil to reduce its resistance or slow down the pulse rate to the coil by increasing the value of capacitor C1.

I always like to build the circuit on a prototyping breadboard as shown in Figure 5-22 before I solder any components or wiring for permanent installation. In Figure 5-22 I tested a few different coils and magnets that I found in my scrap bin to see which ones gave the best sound for this device,

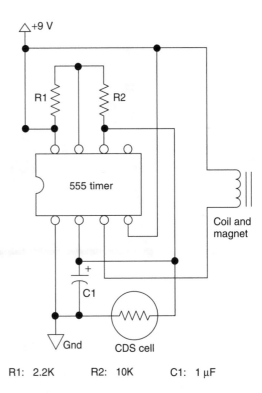

R1: 2.2K   R2: 10K   C1: 1 μF

Figure 5-21   *Light-activated coil-pulsing schematic*

but in the end, the hand wound coils worked the best overall.

With the circuit working the way I wanted, the few semiconductors and coil were transferred to a small piece of perforated board for installation into the usual black box (Figure 5-23). The best way to mount the coil and magnet for optimal noise making is simply to glue the coil to the base of whatever cabinet you plan to put things into and then glue a cap (pop-bottle lid) over the coil to keep the magnet from jumping out of the coil center. The CDS cell should be placed over an opening in the cabinet, if it is not translucent, so the light-activation circuit will function properly. When there is enough light in the room that you can see from one side to the other, the unit should not be making any noise at all. To adjust the sensitivity of the device, just aim the CDS cell to the light source in the room, or to a bright reflecting surface like a white wall or glossy floor.

The unit will not draw much current when the lights are on, so I did not bother installing an on–off switch, and removed the battery until it was

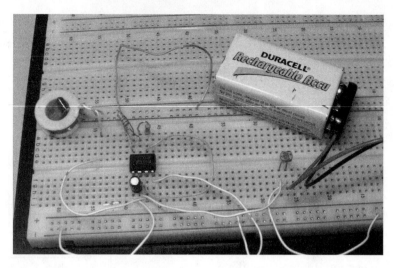

Figure 5-22   *Prototyping the circuit before the final build*

Figure 5-23   *Components ready for the black box*

time to play the prank. The device works great when hidden behind some furniture so it cannot be seen easily. If it is pressed up against a wall, the chewing noises sound as though they are coming from inside the wall due to vibrations. You have to remember to orient the device so that the magnet bounces around on the base of the cabinet. Ensure you label one side as the "top." Now you can tell your pals all about the huge rat or mice you saw scurrying across the floor so they will have something to think about as night falls and the gnawing sounds begin!

Project 20—Footsteps in the Night

This project is very similar to the last one, but instead of rodents chewing in the walls, we will be simulation the sounds of footsteps walking across a hard surface. This device is also night activated, so it can be placed at the end of a long hallway, or at the far side of a room with a hardwood or concrete floor to make it sound like someone is creeping around in the dark. Of course, once the lights are switched on, the sound instantly stops, and the source of the ghostly footsteps cannot be found. To create the sounds of heels or boots

walking along a hard surface, a motor with an unequal length weighted T-bar is placed on the shaft so it will make that well known "click-clop," "click-clop" sound of heels or boots. Feel free to experiment with the size and type of materials used to make the footstep simulator, but for starters, a motor, a few nuts, and a bit of copper wire will be needed as shown in Figure 5-24. This motor was taken from a dismantled CD drive, and the copper wire is a bit of house wiring with the insulation removed.

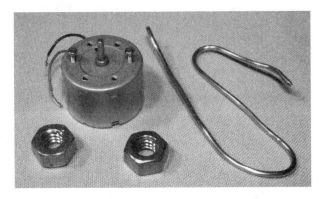

Figure 5-24 *A motor, some copper wire, and two nuts*

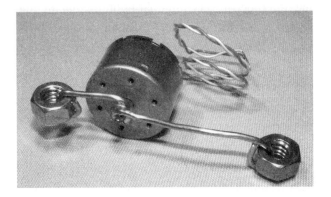

Figure 5-25 *The T-bar and weights mounted to the shaft*

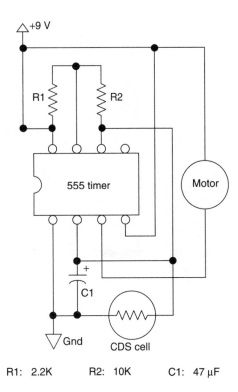

R1: 2.2K R2: 10K C1: 47 µF

Figure 5-26 *Motor-pulsing circuit based on a 555 timer*

Figure 5-25 shows how I made the unequal-length T-bar using the copper wire and nuts to create the walking sounds. The reason the T-bar is longer on one side is so that the weights (nuts) hit the surface of the device at different rates, just like a heel and sole hit the surface of the floor at different rates when a person is walking along a hard surface. The timer circuit will simply engage the motor for a split second, making the first nut strike the surface of the device, and then once the power is removed from the motor, the other nut will drop and strike the surface, creating the beat of walking feet. In my unit, one side of the arm is about 1 inch, and the other is about 2 inches. The center of the copper wire is wrapped around the motor shaft and then held securely with a bit of solder on the wire and motor shaft.

The surface that the weights will strike can be the actual cabinet that everything will be housed in, or it can be something else to simulate whatever type of walking sounds you like. If you have a marble floor, then a tile may be better than a plastic box, or you may want to try different weight objects as well to get a more realistic sounding walk for your environment. The size of the motor can also alter the sound, but I doubt you will be able to power a motor much larger than the type I am using without having to add a driver transistor since the 555 can only sink a small amount of current. Of course, there is no reason you couldn't add a large driver transistor or a relay to engage a huge motor with an actual boot tread mounted to a large T-bar made of wood for some extremely realistic walking sounds. How about a windshield-wiper engine with an actual boot mounted to its arm for a large outdoor unit? Anyhow, Figure 5-26 shows the basic circuitry used to drive the motor directly on the small device I am building. Notice that it is similar to the circuit in the previous project with only a small-value change in capacitor C1 to slow down the pulse rate.

Figure 5-27 *Footstep simulator ready for action*

The motor pulse rate is determined by C1, R1 and R2, so feel free to experiment with the values to get the most realistic walking sound you can. You should also take note of the polarity for your motor because, when the motor is secured to the cabinet, you will want the pulses from the 555 timer to lift the long end of the T-bar, creating the first strike with the weight on the short end of the shaft, allowing the long end to simply drop back down to the surface to create the second strike.

In my device, I just used a bit of hot glue to secure the motor to the outside of the cabinet as shown in Figure 5-27 since there was not enough room inside the box for the circuit board and the battery. Having the motor and weights outside the box also allowed easy alterations of the weights to experiment with different walking sounds, although the simple bolt trick seemed to sound alright at a distance.

The CDS cell was mounted on the outside of the box for optimal light reception, and would shut off the 555 timer in very minimal lighting, making the source of the footsteps very difficult to track down. The device worked best when placed in a large room where the person you are tricking cannot clearly see the device when they flip on the lights. A carport or large badly lit basement is optimal, since the footsteps will echo across the room and make them sound larger than they are. If you want to try some additions to this device, another good idea would be a second 555 timer circuit that only triggers the motor cycle once every few minutes so that the walking does not begin for some time after the lights have gone out. There are many ways that you can modify this simple light-activated timer to trigger all sorts of devices, so use your imagination and haunt your buddies after midnight.

Project 21—Giant Shadow Projector

This project uses a few bright white LEDs to project large crisp shadows on to a surface such as the walls of a dimly lit room or on to curtains from inside a dark room to make ghostly shadows appear from the outside of a house. The concept of this project is very simple, but you can produce some really cool effects if you play around with the position of the highly focused LEDs and the type of shape used to cast the shadows. Project a 10-foot skull on to your picture window for Halloween, cast some shadowy message on the

wall at a party, connect the LEDs to the output of an amplifier and modulate some image to your music, you could even add a few servos and a remote control to make a large shadow appear to move around the room. White LEDs are extremely bright, as you probably know after looking into the end of an LED flashlight and seeing the afterimage for a few seconds. Every few months, a manufacturer will announce even brighter LEDs, and now you can find single units that produce tens of thousands of MCDs. The brightness of an

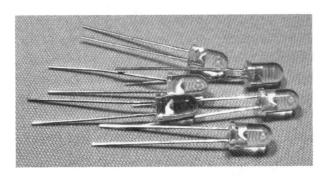

Figure 5-28 *Six very bright white LEDs*

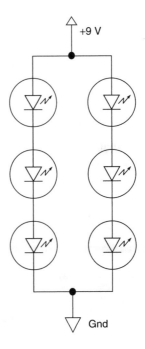

Figure 5-29 *Series and parallel wiring for 9 volts*

LED is measured in millicandela (MCD), or thousandths of a candela. Panel-mounted indicator LEDs are typically in the 50–100 MCD range, where as ultrabright LEDs can reach 15,000 MCD, or higher right up to a million MCD or more for large >1-watt devices used for lighting. For this project, a few highly focused white LEDs rated at 8000–10,000 MCD will be perfect since you only want to cast shadows, not light up an entire room. Figure 5-28 shows six 10,000 MCD white LEDs I plan to use in my shadow-projecting unit.

Depending on the brightness of the LED and size of the shadows you want to project, one LED may be enough, or you may want multiple LEDs, like I have used, to soften the edges of the projected shadows. One bright LED will produce a nice sharp shadow on a wall over an area of 10 foot squared from a distance of about 15 feet, and multiple LEDs will cast several shadows, each an inch or two apart to give a more diffused look to the shadow. Multiple LEDs can also be driven by a single battery by using the appropriate, series, parallel wiring, so no energy will be wasted by a current-limiting resistor, which would be needed in a single LED circuit. The schematic shown in Figure 5-29 is my series and parallel wiring diagram for the six LEDs in my projector, each requiring no more than 3.3 volts. Since three of them are in series, this gives 3 volts to each LED from a 9-volt battery, plenty of light to cast nice large shadows. Once you have a series-connected string of LEDs, you can add as many of those strings in parallel as you like, as long as the battery can source the necessary current.

A typical 9-volt alkaline battery could probably handle 50 or more LEDs, but this would be overkill for projecting shadows and would result in only a few minutes of bright light before the battery began to run down. I recommend that you start with only the minimal number of LEDs you need to create a single series-connected string, then add more later if you find it necessary. Because the leads were plenty long on each LED, I simply drilled the appropriate-sized hole in the mounting cabinet and then soldered the leads together to create the wiring shown in the schematic. Remember, your LEDs may not be the same voltage as mine, so the goal is to wire as many in series as needed so you do not exceed the rated voltage. A little less will be much better than a little more when dealing with LEDs and voltage. Figure 5-30 shows the LEDs wired for 9-volt operation.

When switched on, the six LEDs glow at an amazing brightness that is actually hard to photograph, and painful to look at directly, something that is not recommended, but we feel the need to do anyhow. The little light box shown in Figure 5-31 actually makes a great LED flashlight and can light up a completely dark room with more than enough light to get around safely. When aimed at a lightly painted surface in the dark, the LED projector was great at producing large sharp shadows from any small object placed within a few inches of the LEDs.

Simple shapes that are printed on a standard inkjet or toner printer, and then cut from standard

paper made great shadows when taped to a fine wire and placed a few inches away from the LEDs as shown in Figure 5-32. If a fan is placed near the projected object, it would move around as if coming to life as the fan blew it around on the fine wire that held it in place. Projected against some light curtains at night, I could make it look like

Figure 5-30   *Using the LED leads as wires*

Figure 5-31   *LED light projector completed*

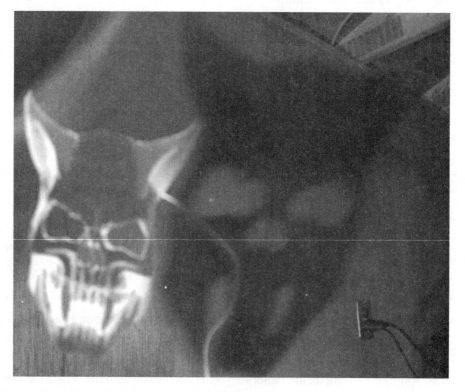

Figure 5-32   *Projecting a large devil skull*

evil entities were moving around the house as viewed from outside, a great Halloween light show indeed. There are hundreds of ways that you can cast interesting or eerie images into a dark room, so use your imagination and whatever junk you might have laying around to experiment.

Certain translucent objects can enhance the image with shades of gray, or even color, and if you have a printer that can print on to transparencies, you can make some very cool shadows that look like very real objects. You may also want to try mounting the light box to a servo motor to animate the shadows, making them come to life as if projected by real objects. At night, you can create a simulation that appears as though people are in the home for a simple home security device. Well, I will leave you to come up with interesting ways to scare your pals when the clock strikes midnight. Hope you aren't afraid of the dark.

In the next section, we will create and summon supernatural powers to bring objects to life!

Chapter 6

Evil Abounds!

Project 22—Voices from the Grave

Here is a device that will magically receive whispers and voices from beyond the grave, or at least it will seem so as you roam around the house listening to the voice fade in and out. The real magic behind this device is an invisible beam of infrared light that carries an audio signal generated by the stealthy transmitter, which can be plugged into any audio source used to generate the ghost sounds or voices. Because the audio signal is carried on a beam of weak infrared light, the received audio will seem to fade in and out with a ghostly static, as if the voice from beyond is trying hard to communicate. The unit will also pick up kinds of bizarre buzzing and crackling sounds from stray light sources, adding to the effect that there is really something out there trying to communicate. The transmitter unit is disguised as a TV remote, since we will use parts from it to generate the infrared light, so it is almost impossible to track down the source of the ghost signal. A commercial example of this device can be found by looking at any wireless headphone system that uses infrared LEDs to transmit the signal instead of radio frequencies. Our device is not nearly as complex as a wireless headphone system, so it suffers signal dropouts and a lot of noise interference from ambient light sources, making it perfect for this project.

You can start with an old TV or VCR remote to get the infrared LED like I did, or simply build the transmitter from scratch using a new infrared LED

and some other hobby case. I thought the TV remote would be a good way to hide the unit, so that it can be placed in any room without looking suspicious as wannabe "ghost busters" try to track down the eerie voices from the grave. If you plan to use a TV or VCR remote control for parts, ensure it has an infrared LED sticking out the end or it may be an RF remote, which will not work for this project. Most remote controls are of the infrared variety, such as the one I am about to dismember in Figure 6-1.

To send audio through the air on an invisible beam of light, we need to modulate the infrared LED with some audio source feed through a low-power audio amplifier. The LM386 1-watt audio amplifier IC is perfect for this job, and will require only a single resistor to limit the LED current and a capacitor to remove any DC voltage from the audio source input. As shown in Figure 6-2, the schematic for the modulated infrared transmitter is ghostly simple. You can feed audio into the transmitter from any MP3 player, computer output or any audio device that has a line output, so your ghost voices can be made on practically any audio appliance.

If you have gutted an old remote control to hide the transmitter parts in, then you may need to remove the original circuit board if there is limited space for the 9-volt battery. You could try the original 3-volt battery source that came with the remote, but you may find that the output from the

Figure 6-1 *A typical infrared remote control*

transmitter is very weak and may only have a range of 10 feet or less. The 9-volt battery is an optimal power source for the LM386 amp and should allow the receiver to work across a large open space, and even around some corners if the wall is able to reflect the invisible infrared light. The 100-ohm resistor can also be swapped for a lower value between 10 and 50 ohms if you want to try pumping up the output from the infrared LED, but try the original value first before you start pushing the upper limits of current capacity for an LED. This circuit is so simple that I didn't bother with a circuit board and just soldered the legs of the components together right on to the top of the original remote control's circuit board. You must cut the traces on the original circuit board where the LED is connected if you plan to build the unit this way. Figure 6-3 shows my lazy method of building the circuit without any circuit board.

If you do the direct solder method, add a bit of hot glue to the components after the unit has been tested so the parts don't shift around once you put the remote control back together. Now your invisible light transmitter is ready for operation, and you can start building the receiver in order to test both parts of this project. Figure 6-4 shows the infrared audio transmitter sending some creepy sounds through the air as they are played into the device through my

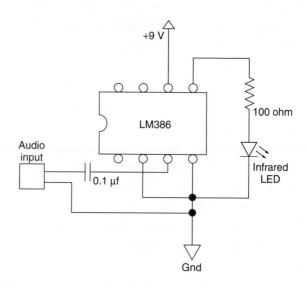

Figure 6-2 *A very simple infrared audio transmitter*

MP3 player. "Carol Ann, come into the light Carol Ann." Yes, you know the routine!

Now let's build the infrared receiver, which strangely enough uses the same LM386 1-watt audio amplifier IC to get most of the work done. Starting with the schematic shown in Figure 6-5, you can see that the LM386 drives a pair of headphones or a small speaker this time rather than an infrared LED like it did in the transmitter. The input now takes the form of an NPN phototransistor, which will create a modulated voltage on the input pin of the audio amplifier as it receives the modulated infrared signal sent from

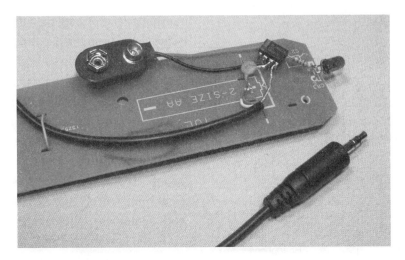

Figure 6-3 *No circuit board. It's ugly but effective*

Figure 6-4 *Audio traveling on an invisible beam of light*

the transmitter unit. Because of the extreme sensitivity of the phototransistor to any light changes, it will also pick up all ambient light sources, resulting in some ghostly yet cool sounds from AC lights, arc lights or even video screens. The ghost voices being sent from the transmitter will be the loudest sounds heard as long as the transmitter and receiver are pointed in each other's general direction at a distance of less than 30 feet. The 10 μF capacitor can be removed from the circuit or added to a switch if you find that the receiver's output is too loud, as this controls the gain of the amplifier.

The receiver circuit has a few more components than the transmitter, and warrants a circuit board and a proper cabinet for the output jack, power switch, battery and a volume control if you choose to add one. A 50K variable resistor in series with one of the headphone wires will become a volume control, if you would like the ability to lower the audio output while using the device. You could also drive a small speaker directly with the LM386, which is why the output jack is a good idea, and makes the device more fun when showing several people how haunted your house really is. The only other requirement is that the photo transistor window be mounted through a hole in the cabinet so you can aim it towards the transmitter or some other light source while you are snooping around the room looking for sounds. With the receiver completed like my unit shown in Figure 6-6, you should be able to apply the power

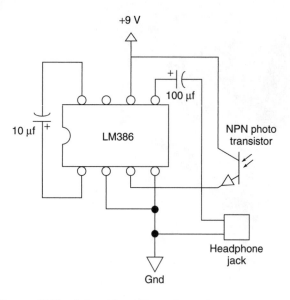

Figure 6-5 *Infrared receiver schematic*

source and hear all sorts of pops, and humming noises as you point the photo transistor towards different light sources in your room. TV screens will be exceptionally noisy as well as any lights that are on dimmers, so you can tell your guests that the ghosts are trying to break through the fourth dimension when you hear the strange buzzing sounds.

The receiver should be tested together with the transmitter by inputting some audio source into the transmitter and placing the infrared LED a few inches away from the phototransistor. If both units are working, the audio source will sound loud and clear on the receiver with minimal distortion or noise at this close distance. If the audio is extremely loud or oversaturated, then try lowering the output level on whatever audio source is feeding the transmitter because you are probably overdriving the input. An MP3 player set on volume level 3 is about right. Once you have both parts of this project working properly, put the receiver into a box and test to see how much range you can get in a lit room and then a dark room. You will see that a clear line of sight or light-colored wall will give the strongest reception, but as the distance increases between the transmitter and receiver, other sources of modulated light will start to overpower the signal. Figure 6-7 shows my completed receiver ready to baffle my haunted house guests with all sorts of communications from the "other side."

If you have a computer in a room that will allow the transmitter to cast its light into a large area of your house, then playing back your ghost voices directly from the hard drive may be the best way to go since you can have a very long sound continually looping. For living room use, just mix

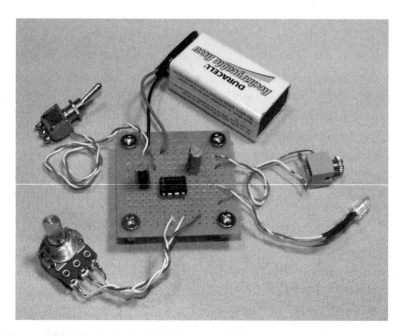

Figure 6-6 *Infrared receiver ready*

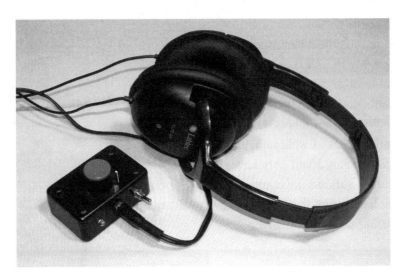

Figure 6-7 *Who ya gonna call?*

down your evil audio track to an MP3 player then use that as a source, although you will have to find a way to either loop the track or press play when nobody is looking. Invite your skeptical guests over and show them the cool ghost voice receiver that you rented from the local exorcist, explaining that you are not sure what to do, which is why you need their help. Start scanning at the far end of the house well out of range from the transmitter so you only hear pops, hisses and ambient noise from nearby light sources. Have them listen as you hear

only noise, explaining that you think the device must be a hoax. As you get closer to the transmitter, the eerie sounds in your recording will start to fade in and out, which should certainly wake up your non-believers in a hurry. When they tell you they heard something, have a listen but pretend not to hear anything. Now they can run around the house thinking the ghosts are only communicating through them as long as you can keep a straight face. If you hear real dead people with this device, please do not call me—ever!

Project 23—Evil-Possessed Doll

Remote control (RC) servos add proportional mechanical remote control to many remote-controlled vehicles such as race cars, boats and aircraft. Proportional control means that the servo's output shaft moves proportionally with the remote-control joystick so you can have fine motor control over whatever the servo may be connect to. For a ground vehicle, this will allow fine control over steering and acceleration rather extreme speed fluctuations often found on very cheap RC toys. For this reason, RC servos are often used in robotics, animatronics and many movie props that need to be controlled remotely. Do an Internet

search for animatronics or humanoid robot, and you will find many examples of amazing animatronic robot faces or multi-legged walking robots based on nothing more than a few servos and some circuitry to move them. A four-channel RC remote-control system that includes the remote joystick, RC receiver and four or more servos can be purchased from most hobby shops for under $100, and will make a great addition to your hobby parts arsenal, especially if you are interested in androids, animatronics or robotics. This project is only a simple example of how easy it is to make things come to life using RC servos and a remote control, but it sure can startle an unsuspecting bystander when an inanimate object begins to move around on its own. My RC set is an aircraft type with a six-channel receiver and a joystick with two dual-axis controllers. I often use this setup to control my long-range spy robots, but it is perfectly suited for animatronics as well, although in my simple possessed doll gag, I will only have two servos in use. Figure 6-8 shows my common RC setup including the transmitter, receiver, battery pack and two of the six standard size RC servos.

Standard RC servos can move a few pounds of weight, which is pretty impressive considering their size. In animatronics, you will often want to move much smaller objects, such as plastic eyeballs, a plastic jaw, or create subtle facial expressions by pushing and pulling small levers glued to the inside of a rubber or latex facial mask. Some extremely realistic humanoids have been made using a dozen servos, a latex mask and some type of computer program to sync the motion of the servos to a

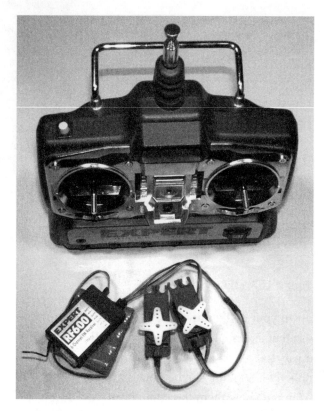

Figure 6-8 *A typical hobby RC joystick and servo setup*

Figure 6-9 *Soon this doll will have revenge*

calculated series of movements. My possessed doll is certainly nothing that looks at all realistic, but it can look around the room and then make thrashing arm movements while welding a deadly plastic butter knife! Like many cheesy slasher movies, this prop is both creepy and banal at the same time. To start with, I found an inexpensive plastic doll that looked as harmless as possible in order to make this prank more humorous. My original plan was to use

one of those classic rag dolls, but dude, those things already creep me out, and there is something truly devious about those pitch black eyes, the way they watch me, silently plotting my inevitable destruction. Umm, anyhow, I pulled out the arms and head out of the plastic doll (Figure 6-9) to retrofit the joints with a pair of servo motors to move an arm up and down and the head left to right.

You should have no problem directly attaching the arms to the servo mounting plates as long as they are not extremely heavy or as large as life-sized mannequin limbs. A typical servo could easily move the arms and head of most plastic dolls or stuffed animals, so it's just a matter of finding the best way to attach the servo body and its mounting plate. As shown in Figure 6-10, I cut the back out of the body to install the arm servo, battery pack and receiver, but did not have enough room for the head servo, so it was mounted inside the head. For non-permanent yet sturdy mounting of the servo parts, a bead of hot glue will usually do the trick without any damage to the body or servo parts. If you want to reclaim the servos for another project, the glue will let go with a little tugging and can be picked away from the servo parts without any trouble. The servo mounting plates are available in multi-packs of all sizes and shapes, so just experiment a little to see which ones fit into the limbs with the least amount of effort. The round servo plate shown in Figure 6-10 fits perfectly into the arm once a bit of the plastic was cut away.

The doll's head was the heaviest part, and although the servo could have handled a lot more weight than that, I decided to install the mounting plate using woodscrews just to make sure the doll did not self-decapitate when threatening my paranoid friends. After installing the head (Figure 6-11), I noticed it was up a little too high for the doll's clothing to hide all of the servo parts. This problem could have been remedied with a little more body modification, but for this simple prank, I just left it as is. Luckily, there was just enough room inside the doll body for the battery pack and transmitter, so no visible wires would be hanging out, making the doll look somewhat normal when it wasn't thrashing about.

With the animatronics ready for action, I added a little black scarf around the doll's neck to hide the

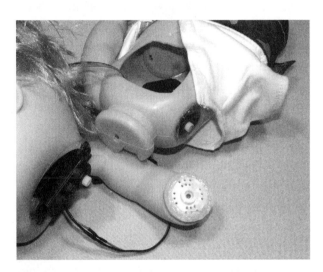

Figure 6-10   *Not exactly Terminator, but a robot nonetheless*

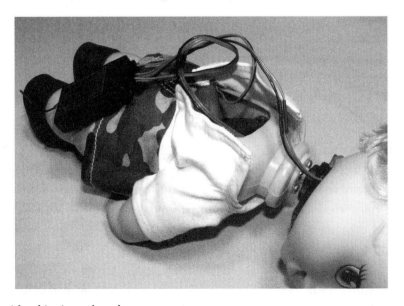

Figure 6-11   *A doll with a bionic neck and arm*

servo mounting plate and then got silly with my black marker and a plastic butter knife. This once innocent-looking doll had a twisted, evil look and wielded a plastic knife in the most threatening manner. Now I could place the doll on a shelf and keep the remote control nearby to bring it to life when it was least expected. Sure, an evil-looking baby doll wielding a plastic butter knife is indeed twisted, but if you've seen the things that I have in my lab, then you would understand that the doll was one of the more normal items in the room, and went unnoticed for the most part. Of course, when a person notices the doll and goes in for a closer look, a few flips of the joystick makes him or her jump as the doll begins to look back and forth while thrashing the knife up and down. A gag like this might be a good implant for a night of ghost stories or a séance, if that is your bag. Have a friend hide in another room and only make small and slow movements on the stick, so the person you are pranking is not sure if they are losing their sanity. When they tell you what the seen, have the doll stop moving until you are not looking, then taunt them some more with your evil-possessed doll. Oh yes, the more evil the prank, the more fun it is!

If you want to take the whole RC servo-hacking project to the next level, search the Internet for robot and animatronics projects done by others and

Figure 6-12 *Move over, Chucky. Evil has a new look!*

you will see many interesting ways to mount, modify and control these inexpensive and extremely useful servos. If you are good with sculpting or mask making, your animatronic project could really dive into the "uncanny valley" with a few servos controlling facial expressions or small limbs. Just remember, don't ever build a robot or android that you can't outrun.

Project 24—Telephone Devil Voice

There are many cool computer programs available that will let you speak into a microphone and change your voice in all sorts of weird and wacky ways. You can turn a man's voice into a woman's, sound like a chipmunk or a demon, or even make robot-like voices like those Cylon raiders from the movie 'BattleStar Galactica'. Some of these voice changers are so powerful that law enforcement use them to disguise voices in ways that are not even noticeable,

while others are just plain whacky and fun. If you type "voice changer" into Google, then you will find many computer-based voice changers available, many of them free or available as a time-limited demo. At the time of writing this book, one of the most widely used voice changers available for download is called AV VCS, a voice changer with many cool features such as robot voice, pitch bending and even gender changing. I call this project

"Telephone devil voice" because I had the most fun making prank calls to my buddies while using the software to change my voice to sound like the dark lord himself. There is just something creepy about picking up the phone at 3 am and having the devil proclaim that you must now pay in full the contract that you signed in order to become a rock star (or so I am told). Of course, you could also crank up the pitch to sound like a twisted gremlin, or use some of the more subtle gender-altering controls to morph your voice into an unrecognizable yet believable character to pull one over on your unsuspecting pals. Regardless of your intentions, you must first obtain some voice-changing software and play around with it to see what it can do. Figure 6-13 shows some of the controls on the VCS voice-changing software running on my computer.

Now, you could just crank up your computer speakers and hold the phone up to them while you speak into the microphone, but this would not only sound terrible, but most likely end up creating a lot of unwanted feedback, so we will create a very versatile little box that will allow you to interface just about any audio source into the telephone. A typical telephone system uses a simple two-wire cable to send and receive audio, but there is nothing in common between a telephone system and a typical audio patch cord, so care must be taken when attempting to interface the telephone to any type of audio equipment. When the phone is on the hook, there will be 40–50 volts DC presented across the wires and, when it rings, this voltage will peak at 90–100 volts AC, which is more than enough voltage to wake you up if you happen to be stripping wires with your teeth while your circuit is live (not recommended)! The current available on the telephone line is minimal, but still, neither your body nor your sensitive electronic devices will appreciate a direct connection to 100 volts. When the phone is off the hook, or in use, the voltage drops from 50 volts to somewhere between 5 and 15 volts depending on how many devices there are

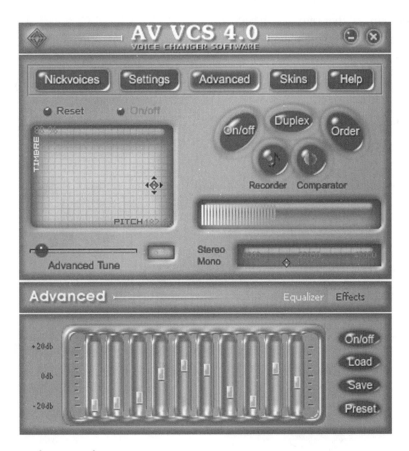

Figure 6-13 *VCS voice-changer software*

connected to that line. Obviously, we are going to need some type of isolation to send or receive audio from this hostile pair of wires. This simple device can connect the phone line to practically any audio device to either playback audio into the phone line, or record the audio from the phone line, so it has many uses beyond the one presented in the project. Before you get out the soldering iron, you should know that there are only two wires used in a typical single-line residential phone cable, although the cable will most likely have four wires. If you follow any of the phone boxes wiring back to the main terminal, you will notice they all connect to a main block using only two of the four wires—a red one and a green one (yellow and black are not used). The green wire is specified as "tip," and the red wire is specified as "ring," and although this polarity is very important in most telephone equipment, it means nothing in our simple interface, as it is non-polarized, meaning you can connect it either way to the phone system and it will work the same. Take a look at the schematic in Figure 6-14, and you will see that only five basic components are needed for the telephone audio interface—three 0.1 µF ceramic capacitors, a 1:1 audio transformer and a variable resistor.

The capacitors remove any DC from the telephone line and allow for "invisible operation" on the phone system. Invisible operation means that the device does not load down the phone line at all, so it will not be detected as an in-use extension or cause the phone to go off the hook. The 1:1 audio transformer further isolates your equipment from the phone line by electromagnetically coupling the two devices together. The variable resistor is used to control the input–output level to the audio transformer just in case your source device can't do this. All of these devices can be salvaged from practically any defunct telephony device such as a modem, fax machine, answering machine and, of course, a telephone. As shown in Figure 6-15, a

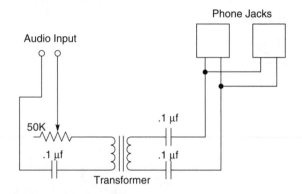

Figure 6-14 *Schematic for the telephone audio interface*

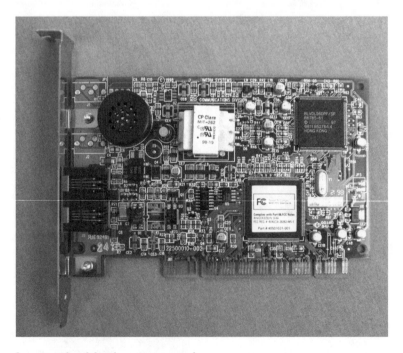

Figure 6-15 *56K modem sacrificed for the greater good*

prehistoric 56K modem is sacrificed for this project, as it contains all of the capacitors, the dual phone-line jack and the audio transformer (large block in the center) that will be needed.

The audio transformer is easy to identify, as it will be the largest component that looks like a block of approximately 1 inch squared, with two or three terminals at each side. Unsolder the transformer and measure the impedance across the two terminals at each side of the device. If there are three wires, ignore the center one. The impedance at each end of the transformer should be equal, since the number of turns in each winding is the same, which is why it is called a "1:1 transformer." The dual phone jack is optional, but if you plan to connect this interface to a phone line that already has a phone connected to it, then there will be no need for a Y-adapter, since the dual phone jack will fill that function. The circuit is very simple, so it can be built to a minimal bit of perforated board, or right into a telephone extension box as shown in Figure 6-16. The telephone box installation (top photo) is convenient because you can plug directly into the phone line, and then plug your phone back into the box without needing a Y-adapter.

If you intend to use the audio interface with a computer sound card for either input or output, then you can leave out the variable resistor and simply control the volume level with your computer mixer. To use the device with a software-based voice changer, simply connect the input of the transformer to the output of your sound card and then plug the phone cable into the phone jack in your wall. Listen on the phone while you set the audio level on your computer for a clear and loud level while playing

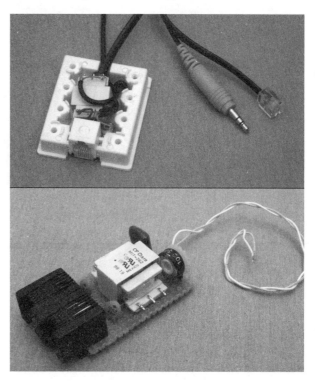

Figure 6-16 *Telephone audio interface shown in two flavors*

back some music or while you speak into the microphone. The other phones sharing the same line should not be affected and free from any buzzing or AC hum when your device is connected to the line. If there is a problem with buzzing, or the phones are off the hook (no dial tone) when your unit is plugged in, then disconnect the interface and re-check the wiring. If everything is working OK, set your voice changer the way you want, and have a great time pretending to be someone or something else while you call your buddies in the wee hours of the night. You can also have the voice changer ready and waiting so the next time you get assaulted by a telemarketer, the joke will be on them as they try to sell their wares to devils, gremlins or robots!

Project 25—Evil Lurching Head

This project is guaranteed to make your victims jump clear out of their shoes and cling to the ceiling! It doesn't matter how tough your buddies

are because when a life-sized head pops out of the least expected place, they will jump or screech every time. This spring loaded head can be placed

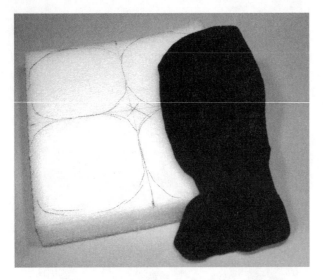

Figure 6-17 *A few foam disks will form a head insert*

Figure 6-18 *Foam disks glued and ready to cut*

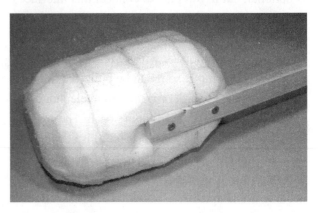

Figure 6-19 *The head is mounted to the swinging arm*

behind furniture, a shrub, building, or outside a window for activation by remote, wire, or light, depending on how you decide to trigger it. Since the concept is very simple and easy to build, you can easily adapt the device to suit your particular application, be it indoors or outside. Feel free to mount whatever you like at the end of the spring-loaded arm, but keep in mind that the heavier the object, the larger the spring that will be needed in order to make the object pop into view quickly. A foam head used to display wigs and sunglasses are perfect and can be covered in masks or even dressed up to look somewhat realistic. I did not have a Styrofoam head at my disposal, so I simply found some packing foam and traced out four 6-inch diameter disks to create my own simple head. My idea was to make a crude head-shaped foam insert to fill the black balaclava as shown in Figure 6-17.

The foam disks were crudely cut and then glued together using some spray adhesive. A talented sculptor with a sharp knife could have carved up a very realistic head from the foam block, but since I was just going to throw the mask over it, a basic human head-sized globe was fine. The foam block is very lightweight, so it should spring up very quickly, and not cause a problem if it is allowed to thump off a window during outdoor attacks. Figure 6-18 shows the four foam disks ready to be cut after the spray adhesive has set.

The arm used to swing the head into place should also be as light as possible, so a wooden stick about the same size as a broom handle will be perfect. The length of the stick will depend on your target installation, so it must be made long enough to place the head over the object that it is hiding behind. For outdoor use behind a window, the length of the arm is not critical since you will have to place the base of the launcher on top of a ladder or tall object to bring it up to the window height anyhow. A 3-foot long arm should be just about right for most furniture if you just want to make an all-purpose lurching head. The arm should be fastened to the head using a few long woodscrews and some glue as shown in Figure 6-19. If you are covering up the foam, a little duct tape around the head and arm will hold it very securely, and there will be no chance of a spontaneous decapitation when the head swings into place. Figure 6-19 also shows the foam head after chopping away the corners for a little more human-like appearance.

The launching mechanism is very simple, consisting of a small hinge and a 6-inch long spring from the hardware store. The spring is about the size of a large marker, and can be pulled apart by hand without needing a great deal of force. If the spring is too strong, your lurching head will become a cranium-launching trebuchet, which would certainly be cool, but not the goal of this project. If you can't

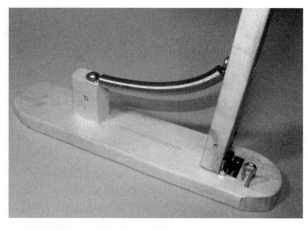

Figure 6-20 *The spring-loaded arm and mounting base*

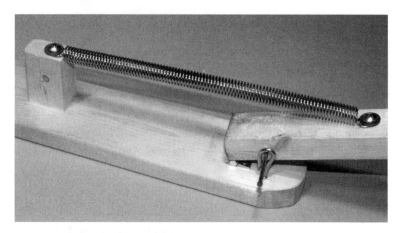

Figure 6-21 *Loaded and locked by the finishing nail*

find an appropriate spring, then a few large elastic bands or even a chopped up bicycle inner tube could fit the bill as well, but you will have to experiment to get the perfect amount of force and speed from the swinging arm. Figure 6-20 shows how the swinging arm is mounted to a wooden base that allows it to be pulled into the upright position by the spring. Also shown is the small eye hook that allows the device to be locked into the loaded position by placing a finishing nail through the hole of the eye hook into a hole in the arm.

Figure 6-21 shows the arm in the loaded position, held in place by the finishing nail, which can be easily pulled out to send the arm flying back into the upright position. The actual trigger mechanism can take many forms from a simple thread that is manually pulled to a remote-controlled servo system like the one presented in

the next project. Another good mechanical launch system that uses an electromagnetic solenoid was shown in Chapter 3, Projects 6 and 7, if you want to automate the lurching head. The nail or pin that releases the arm should come free with almost no friction, so make sure the hole in the wooden arm is slightly bigger than the actual pin.

The base of my lurching head unit is quite small, so I have to weigh it down with a few pounds when I set up the device or the action of the swinging head makes it fall over. In outdoor installations, it can be tied down to the top of a ladder or bucket so the head will pop up smack dab in the center of a window, scaring the wits out of the person inside as it thuds against the glass. The foam-filled balaclava does not look all that realistic on second glance, but it certainly gets the initial shock factor prize. In Figure 6-22, the

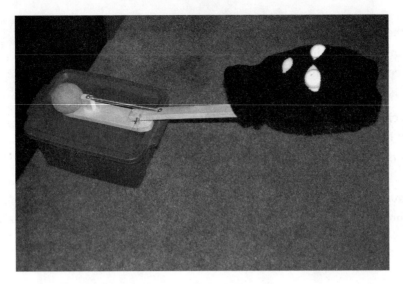

Figure 6-22 *Waiting silently for the right moment*

ping-pong ball-eyed lurching head awaits its next victim, patiently waiting behind a filing cabinet for the next person to open the door and pull the launch pin. Notice the three-pound rock I used to hold down the base to keep the head from jumping off the platform right over the cabinet. In this installation, a length of sewing thread was connected to the door to the room so that opening a cabinet drawer simply pulled the launching pin.

The lurching head can also be triggered by a light-activated solenoid like the one shown in Chapter 3, Project 6, or by the remote-controlled servo system shown in the next project for an all-out haunted house setup. Fishing line is also great for triggering devices in dimly lit rooms or outdoor settings because it is extremely difficult to see and can be strung hundreds of feet without any problem. The most effective ways to get someone with the lurching head is to have them trigger the launching mechanism on entering a room by direct connection to a door or light switch, or by manually launching the unit with a hidden string at the opportune moment for maximum hair-raising impact. Having the head pop up in front of a window at night after telling your buddy that some twisted evil-looking figure just hobbled across the yard would certainly be a good setup, especially after watching four hours of gory movies!

Project 26—Give Us a Sign!

The next time you are sitting around a candle-lit room in dark robes with your pals trying to conjure up spirits from the nether world, make them think you have really made contact using this funny prank. This remote-controlled mechanical lever system is really a multi-purpose device for secretly moving objects, knocking on walls, flipping on and off lights, or triggering other mechanical devices. With the ability to move four or more levers covertly, each capable of pulling a few pounds, there is no end to the ways you can creep out your friends in your haunted house or room. In this project, I will use the device to prove that I have made contact with ghostly spirits at a séance by calling out "give us a sign," and receiving a few knocks on the walls just as the lights mysteriously go out all by themselves. Of course, all this magic happens as I secretly move the joystick hiding under my chair or

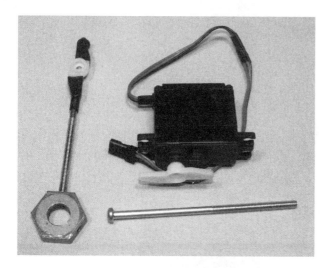

Figure 6-23 *Heat-shrink mounted bolts*

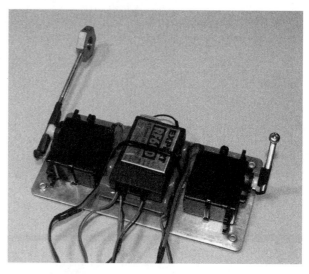

Figure 6-24 *Everything is mounted on to an aluminum plate*

on the floor while nobody is looking, which activates two of the servos I have on my four-channel receiver. The same commonly available RC servo and transmitter set used earlier in this chapter are at work again, but this time as levers that pull and push to create a mechanical reaction on some other device. Shown in Figure 6-23 is the simple method of heat shrinking a long bolt to one of the plastic servo plates that come with the servo motors.

The bolt in the left of the photo has been heat shrinked to one of the servo plates and carries a metal weight at the end so it can knock against walls or wooden objects as the servo is activated slightly. An effective wall or door knocker is made using a 4- or 5-inch long bolt with a large nut or marble tapped to the end as a weight that will strike the surface. The servo only needs about ¼ inch of travel in order to make an effective knock, so there is not much servo noise or lag time as you activate the joystick. My setup is currently using only two of the six available RC receiver channels, one to knock on the wall, and the other to pull down on the light switch, but I could easily add more servos as I come up with haunting ideas to freak people out. As shown in Figure 6-24, the dial servos, RC receiver and battery pack are simply tie-wrapped to a small aluminum plate (box lid), which can be tacked to a wall under a light switch out of view.

The flipping of the light switch using a hook and thread is pretty crude, but it does allow instant and easy installation just about anywhere by simply tacking the unit to a wall. The servo can also be made to turn on a light dimmer by connecting the shaft to the dimmer knob, but this would be easy to spot if the light switch was in full view of your guests. Clear fishing line or light-colored thread hides very well, but in order to make a good photo, I used dark thread for my photo of the light switch hook shown in Figure 6-25. The thread is set tight while the bolt connected to the servo is at 90 degrees to the wall so that it will pull down on the light switch as soon as it is moved down in the same direction. It may take a bit of tweaking to get the right thread tension, but this is easily done by wrapping the thread around the bolt to make it tighter, or by repositioning the device on the wall.

The device is shown held to the wall by two thumb tacks in Figure 6-26. Placing the device as close to the floor as possible will help keep it out of view and, if needed, a cover that looks like a wastepaper basket or some other object that does not look out of place can be used. The RC receiver has a very long range, so no antenna will be needed on the receiver, and the remote control antenna can be collapsed all the way down without any problem. With a little ambient noise or music in the room, the servo noises will probably not be noticeable as the knocker or light switcher is activated, but if you think they may be too loud, then install a noise

Figure 6-25 *The servo can now pull the light switch off*

baffle made out of cardboard and tissue paper over the servos to keep them silent.

Well, there you go, now you have a multipurpose remote-controlled room haunting device ready to switch lights, knock on walls, yank pictures off the wall, move objects, drop objects or trigger other evil contraptions. The remote control can be mounted under your chair using a bit of Velcro® tape for easy access or simply placed under a table so you can control the stick with your feet. Because the action will take place on the other side of a room or at some other distant location, you

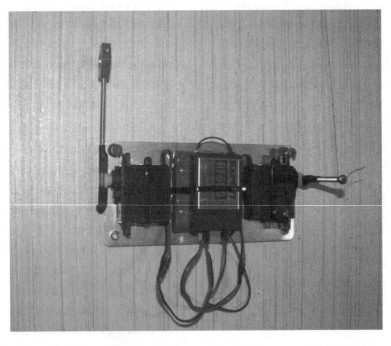

Figure 6-26 *Wall knocker and light switcher working*

will not be the center of attention, so moving the joystick will be easy to do without being noticed. Using all six servo channels, you could rig up an entire haunted house and have your paranoid guests running out the door as you awaken mysterious forces from the other side.

Project 27—Flying Ouija Board

I can't count how many times I have heard silly stories of Ouija boards flying across the room or spelling out evil phrases without any human interaction. Now that the world is connected via the Internet, I dare you to type in the search phrase "Ouija dangers" and spend a few minutes reading some of the warnings posted by superstitious people who think that these simple toys are like gateways to a dark universe full of angry spirits. No really, I highly recommend you spend a few minutes reading about the all-powerful Ouija board before you dare touch one; it's for your own safety—doh! OK then, you are still reading, which proves that you are not afraid to unleash angry demons into your living room by moving a bit of plastic around a 12-inch piece of cardboard that you bought in the toy department, so let's begin. You will obviously need a Ouija board like the one shown in Figure 6-27 in order to pull off this prank. The board is about the same as any board game and comes with a plastic device called a

planchette, which is used to reveal the message by pointing to letters or words on the board.

The general idea is that answers come from the spirit world, or as some believe, the "dark side." Whoa, I'm getting goosebumps! Some believe that the messages are spelled out due to something called the "ideomotor effect," a psychological phenomenon wherein a subject makes motions unconsciously. Personally, I think it's just hocus pocus, like every other ghost story, and people just like to pass them on. This time, however, there will be no second guesses. When that planchette stands up all by itself and starts to twirl, there will be no doubt that evil spirits have manifested themselves into the game, or at least that's what it might look like. The magic we are going to use is called "magnetism," and I guarantee you will not have to wear a black robe and boil up a cauldron of newt tongues in order to make it work! There is a bit of danger involved though, since you are going to be handling extremely powerful NIB (neodymium)

Figure 6-27 *A classic Ouija board*

Figure 6-28   *A stack of seriously powerful NIB magnets*

magnets, which can really make you sorry if not handled with extreme caution. If you have never had the chance to work with large NIB magnets, then let me warn you that they are like no other magnet you have ever played with, not even close. Each magnet in the stack shown in Figure 6-28 measures 1 by 3 inches, and can easily lift a large filing cabinet by itself. It takes a great deal of force to twist one of the magnets away from the stack, and must be done wearing gloves in order to avoid having a finger pinched if one of the magnets slips while handling it.

I have had the joyous opportunity to experience first hand what it feels like to have a finger slammed between two of these magnets, and I can tell you that I now have a new respect of them! Yes, a purple fingernail for 3 months is a clear reminder that these things can really be dangerous. Now that I have warned you, just like the Internet store where you might purchase them from will, we can move ahead and get back to the project. You will need at least one large NIB magnet, and either a few smaller ones to hide under the original Ouija board's planchette, or another large NIB magnet to hide inside a home-brew planchette. The large magnet will be worn strapped to your leg so you can secretly manipulate the planchette (which will contain another magnet, or several smaller ones), making it move to certain chosen board locations, or make it jerk wildly, flipping around or completely off the board. Unlike normal magnets,

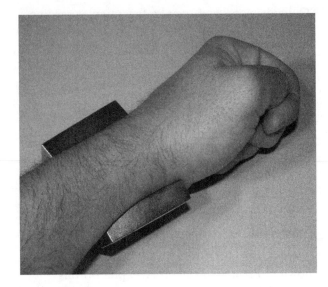

Figure 6-29   *Foolish magnet tricks, not recommended*

which have a moderate amount of pull for an inch or two, the large NIB magnets are extremely powerful up to 6 inches or more and can move large objects through solid table tops with ease. In Figure 6-29, some foolish dude is demonstrating how easily the magnets clamp against his wrist as he tries to shake them free without success by flailing his arm up and down. Yes, like I said before, they are crazy powerful, so keep them wrapped in cloth, or at opposite sides of your workbench when they are in storage.

If you want to use the original Ouija planchette, then you could install a few small NIB magnets under the plastic, or right into one of the three small tubes that hold the felt legs in place. Figure 6-28 shows two smaller cylindrical NIB magnets stuck to the pile which would be perfect for mounting directly under the original planchette using a bit of hot glue or tape. The only disadvantage of this method is that it will be difficult to hide a large magnet under the plastic planchette, and a really small magnet will not let you make any wild movements or flying effects. I chose to fashion a crude planchette from a slab of wood I found in my garage that was large enough to conceal the 1 by 3 inch NIB magnet right inside the body so it would not be seen if a suspicious person were to flip it over for observation. I traced the original shape on

Figure 6-31 *Magnet installed and hidden under felt*

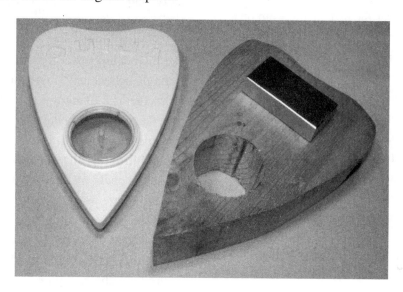

Figure 6-30 *Cutting a new planchette*

to the wood and then cut the new planchette with a jigsaw and drilled, like the original plastic design. The thick wood is better than the plastic for this project because it will easily survive a journey to the floor as it flies off the table if you really want to wake up your séance participants. Figure 6-30 shows the initial planchette cut from a bit of scrap 2 by 6 inch board. The viewing window was also made so that the small plastic disk from the original planchette could be installed.

Using a hammer and a flat-head screwdriver as a chisel (what else are they good for), I gouged out a rectangular hole into the new planchette that would hide the NIB magnet. The magnet was then sealed with some hot glue just to be safe before adding

the felt lining that would not only conceal the magnet but allow a smooth, friction free movement over the board. Consider adding a thin piece of wood or plastic glued over the magnet in the shape of the planchette to further seal the magnet in place, just to be safe. Spend a bit of time sanding the rough wood and paint or varnish the new planchette to make it look like it is actually part of the original game. The viewing window can be popped out of the original plastic planchette and installed in your home brew unit to further make it look like the real deal. Figure 6-31 shows the rough-cut hole, which now hides the magnet right before covering the bottom of the planchette with black felt.

To manipulate the new magnetic planchette, strap one or more large NIB magnets to your leg so they are under your pant leg just above the knee. This allows you to covertly move the planchette to any place you like by lifting your leg so the two magnets are less than 6 inches apart. If you have the leg magnet set to repel the planchette magnet, the planchette will move to a random place on the board by lifting your leg so the magnets are about 5-6 inches apart. If you come a lot closer, then the planchette will move very quickly away from the repelling magnet, and may even fly right off the table, or flip around and stand on its side, for a truly evil-looking effect. To increase control over the planchette, strap on the leg magnet so it attracts the planchette magnet, then it will follow your knee with precise movements so that you can spell out words. If you come very close to the table top with your knee, you can make the planchette feel as though it were glued to the table top, creating an eerie feeling for those that are trying to move it. An old camera or laptop bag strap makes a good magnet strap if you glue one side of the strap to the magnets to hold them in place. You can also use a belt or dog collar to hold the magnets to your leg, just keep in mind that the strap should not cut off your circulation nor should it allow the magnets to come free from

the belt. Figure 6-32 shows the camera bag strap I glued to the backs of the magnets, which I also sealed with duck tape.

With the NIB magnets strapped to your leg just above your knee, practice moving your leg around to manipulate the new magnetic planchette. You will want to mark the belt magnets "repel" or "attract" once you figure out which side does what so you can either control or repel the planchette. The repelling motion seems to be the most dramatic effect, and it can cause the planchette to literally stand on end and spin depending on how fast or close you move the leg magnet. If you really want to mess with your superstitious séance participants, then the magnetic attraction system will be more fun as you get to spell out words and take total control over the direction and position of the planchette any time you like. Another thing to practice is looking innocent while you move your magnetic leg, using as little upper body movement as you can. If the table top is low enough, lift off with your toes so your body sits still in the chair, and nobody will ever see you do it. With a little practice, you will have your Ouija board spelling out complete words with ease, and the next time your buddies feel like telling ghost stories, they will be able to do it without lying. Figure 6-33 shows my new magnet planchette

Figure 6-32 *Installing the magnets to the leg belt*

Figure 6-33 *The Evil spirits are angry tonight!*

flying into action just before standing on end and spinning around like the dark lord himself had control over it. You'd better call a priest to do an exorcism!

All of these projects will certainly help you conjure up some evil spirits to fool your friends, but I'm sure that a true Evil Genius like yourself can dream up a host of new devices that will bring the "dark side" to life for that perfect gag to unleash on the superstitious to whom you owe one. Evil is abound! In the next section, we will inflict some harmless, but effective "shock and awe" projects to bring your evil deeds to a new level.

Chapter 7

Shock and Awe!

Project 28—The Barbeque Box

Shock your buddies silly with a simple barbeque starter? Who would have thought it? Yes, it's true that the little red button that ignites your gas grill can deliver several thousand volts, and will run indefinitely without requiring batteries. It may seem odd that you can generate thousands of volts with a marker lid-sized device that uses no apparent power, yet you can only get a meager 120 volts directly from the outlet on your wall. Of course, the electricity from an outlet is emitted at a lethal amperage level, whereas the high-voltage output from the barbeque igniter contains such little amperage that it is similar to a powerful carpet shock, which is why it is harmless. The igniter generates high-voltage when a spring-loaded hammer smashes against a crystalline material such as quartz, causing a charge separation to occur (piezoelectric effect). If you have built some of the audio projects in this book, then you will be familiar with the "piezoelectric effect," since this is the same process (in reverse) that makes the piezo buzzer vibrate when a voltage is applied. These piezoelectric igniters can also be found in some lighters; however, they are somewhat too small to generate a large enough spark for this purpose.

A piezoelectric igniter is easy to identify as shown in Figure 7-1, since it will be nothing more than a round tube with a large red button at one end.

To remove the igniter from the barbeque, squeeze the two wings together and pull it from the hole in the sheet metal frame where it lives. The wire at the end will simply pull off to expose the high-voltage output terminal. Since the igniter is easy to remove from the barbeque, it can be reinstalled once you are done using it for your evil bidding, so you won't have to singe off your eyebrows when summer comes and you feel like firing up the old gas grill. Of course, you can always find replacement igniters from any store that sells gas grills, as they are pretty universal in design. Feel free to test your igniter's output by holding the two terminals while you pop the button down in order to smash the crystal and release the angry electrons into your fingers. You will feel a sharp pain much like a very large carpet shock delivered after 10 minutes of rubbing your wool slippers against the carpet on a nice dry day. You will discover that the shock isn't enough to make you cling to the ceiling, but it does wake you up, especially if it wasn't expected, which is the key to this prank. To effectively deliver the voltage to the victim's fingers and help disguise the igniter, we will need to connect two wires to the device—one to the output terminal at the end, and the other to the ground wire as shown in Figure 7-2.

It may be difficult to solder the wires to the terminals since they are not the type of material that solder will stick to, so you can simply wrap the bare end of the wire around the contacts, then secure them in place with tape or hot glue. The output voltage is so high that even a bad

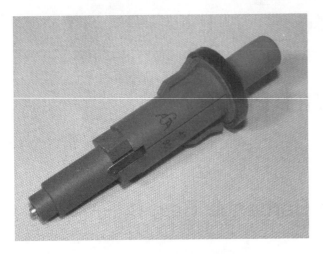

Figure 7-1 *A piezoelectric barbeque igniter*

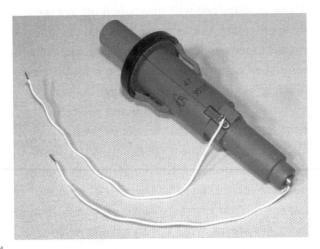

Figure 7-2 *Connecting wires to the two terminals*

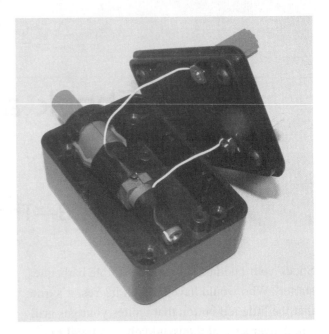

Figure 7-3 *Installing the igniter in a case*

Figure 7-4 *Need your pulse checked today?*

connection between the output terminal and connecting wires will result in voltage travel, since voltage at this level can jump across small gaps with ease. With your igniter connected to the output wires, your job will now be to concoct creative ways to disguise the device so that the high-voltage discharge is least expected, which always seems to make it more of a "shock." My favorite system is some type of generic-looking black box that I can stick a label on to convey some useful purpose other than blasting a few thousand volts into an unsuspecting hand. However you choose to conceal the device, ensure that the two output wires do not cross or come within an inch of each other. Even though there is insulation on the wire, the high voltage will easily jump across a gap of up to an inch, ignoring all the rules

of lower voltage wiring. As seen in Figure 7-3, I installed the igniter in a small black box with a pair of terminals placed far enough away from each other that there is no chance of a spark jumping the gap.

The terminals can be bolts, banana plugs, or just about any metal object that your friends will not be afraid to grab on to. If you are a constant practical joker, then a box with two bare wires hanging out of the end is certainly going to scare away anyone who knows your Evil Genius ways, so try to be as subtle as possible in design. I chose to disguise my barbeque box as a pulse checker that supposedly measures your cardio fitness level once you place you fingers on the end of each terminal. I like to wait a few seconds before hitting the button so I can ask the unsuspecting person holding the device if he or she can hear the beep yet. Once your victim is concentrating for the audible beeps that never occur, let those beautiful volts fly, and once again deliver another successful Evil Genius prank. Figure 7-4 shows my final design, which is so simple that it hurts (literally).

Well, there you have it, a few thousand volts without any electronics or even a battery! Of course, after a while you can snap that little red button all day and not even twitch as you get used to the extremely low amperage shock delivered. If you want to light up your life a little bit more, then read on, I will show you a few more ways to make the sparks fly with a lot more power.

Project 29—Simple Induction Shocker

In 1831, Michael Faraday found that the electromotive force (EMF) produced around a closed path is proportional to the rate of change of the magnetic flux through any surface bounded by that path. Of course, that description of induction is a bit "nerdy" for my liking, so let me put it to you in a context that much better suits this book. If you throw a 9-volt battery across the windings of a transformer, you will get a shock if your fingers are touching the wires when the power is released. This induction effect is a bit strange, since the high voltage is only delivered as the power is removed from the windings, but this is due to the way electromagnets work, and great for creating fast shock pulses that feel a lot stronger than the one produced by the piezoelectric igniter presented in the previous project. For this project, you will need a small appliance transformer and a 9-volt battery as shown in Figure 7-5.

The exact size and number of wires protruding from the transformer are not really important; as long as you rip one of these from a small appliance, it's guaranteed to do the job. A great source of these small transformers is old boom boxes (also known as "ghetto blasters"), AC wall adapters, video appliances, coffee makers with timers, and so on. The transformer will look similar to the one shown in Figure 7-5 and will have two sets of wires coming from each side of the body. On one side of the transformer (the primary winding), there will be only two wires, which would normally be connected to the 120-volt AC supply. The other side (secondary winding) might have between two and six wires, depending on what voltages the small appliances needed, although three wires on the secondary winding is typical. An AC transformer will change voltage between the primary and secondary windings depending on the ratio of windings from the primary to the secondary.

Figure 7-5 *An AC transformer and a battery*

In small appliances, the primary winding of a transformer is connected to the 120-volt AC line and will contain many more turns than the secondary winding, which is why there is a much smaller voltage produced on the secondary winding. Typically, small appliance transformers will reduce 120 volts to less than 20 volts to power the electronics. All you have to do is identify the primary winding for this project, since it will be used to create a high-voltage induction when connected to the battery. As stated before, the primary winding will be the two wires sticking out of one side of the transformer, and often they are the same color. The schematic diagram for this device is extremely simple as shown in Figure 7-6.

In the schematic, one wire of the primary winding is connected to a battery and then to an output, and the other wire of the primary goes to another output. You might think that this odd circuit may seem like it does nothing, since no current is flowing in the open circuit, and you would be correct. If you take each output terminal (one in each hand) and connect them together to close the circuit, the primary winding will be energized, but still nothing happens at this exact point. It's when you disconnect the two terminals that things become interesting—a nice snapping sound and a good shock will flow through your body. When you open the circuit, the energy stored in the electromagnet collapses back through the now open circuit and sends a high-voltage induction into your fingers. So simple, yet so effective! Figure 7-7 shows the simple wiring needed to connect the battery to the primary coil as indicated in the schematic diagram.

The unused transformer wires should be taped off, but feel free to play around with the other wires to see how the shock effect is different. You will likely find that the primary windings will offer the greatest shock, whereas the secondary windings will make a louder spark, but with much less shock factor. Now, once again you are faced with some creative thinking tasks in order to disguise your device in order to deliver it to your rightly suspicious friends effectively. This project lends itself to hundreds of cool designs, since the

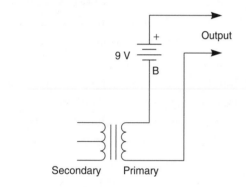

Figure 7-6 *Simple induction shocker schematic*

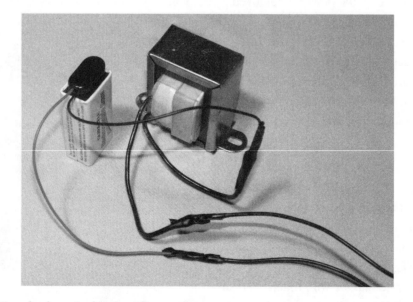

Figure 7-7 *Induction shocker wired and ready*

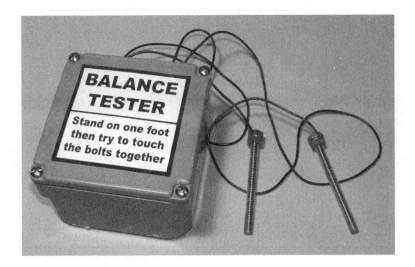

Figure 7-8 *You will always fail this balance test!*

goal will always be to touch the two wires or terminals together. You could wrap the guts into a pair of tinfoil balls and tell you victim that it is almost impossible to touch them together, or put the unit into some type of fake strength-testing device that closes the circuit during use. I came up with the balance testing machine as shown in Figure 7-8, and it is very effective at getting the operator's attention away from the inevitable shock that will soon come.

The balance tester is a great shock-delivery system, since the user is so preoccupied with standing on one foot that they always get caught off guard. "Hey dude, check this impossible balance test out, you have to jump up and down on one foot then try to touch the bolts together while holding out your arms." Of course, when they finally get those bolts together, the joke is on them! This device will run for a very long time on a single battery since the circuit is only closed for a few seconds at most. However, make sure that when the device is not in use that the bolts or wires are not touching each other, or your battery will be drained in a few minutes. Another fun game to play with shocking devices is what I call the "gladiator circle." Have several people joint hands and then the person at the end of each chain has one bolt or terminal each. Complete the circuit to deliver the electric shock, then see who lets go first, so they can be eliminated one at a time until only the last two gladiators are in the circuit. To make the shock more intense, wet your fingers, or move on to the next project.

Project 30—Strong Pulse Shocker

This version of the induction-type shocking device has two advantages over the last unit: it is much more powerful, and it has semiautomatic operation. The output from this device is a lot stronger because we will be using the transformer the way it was intended to operate, just in reverse. The secondary winding will be fed the voltage from your battery, which will create a much higher voltage on the primary when the induction takes place. It's kind of like reversing the operation of the transformer to give back a few hundred volts rather than reducing the voltage. The other advantage to this setup is that the shock can be triggered by a motion-sensitive switch, so you can just let the

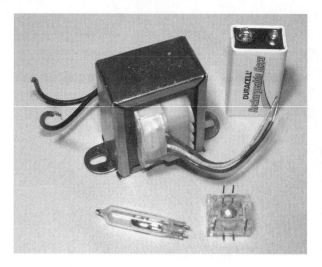

Figure 7-9 *Transformer, battery, and motion switch*

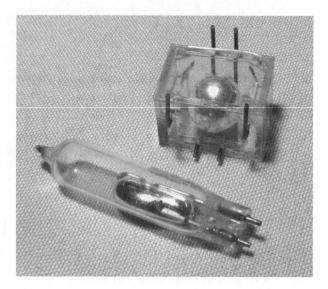

Figure 7-10 *Mercury switch (bottom) and ball switch (top)*

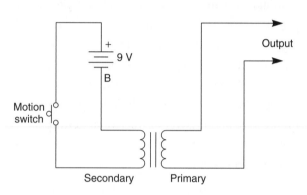

Figure 7-11 *Pulse shocker schematic*

shocker sit around waiting for a curious victim to pick it up and get an unexpected wallop. You will need the same type of transformer as used in the last project, a 9-volt battery and some type of motion-activated switch, such as a mercury or ball switch. The components are shown in Figure 7-9.

A motion switch will close the circuit when it is tilted or shaken, so it is a perfect solution to give this project automatic operation. These switches can be ordered from electronics supply stores or found in older dial-type thermostats and electric heaters with automatic tilt sensors. You can even make your own crude motion switch by hanging a bolt on a spring so that it hits a metal ring if tilted. I find that the mercury switch is the best solution, since they are completely silent, never wear out and can be easily mounted to the enclosure using a tie wrap or bit of tape. Ball switches also work very well but are not quite as sensitive to motion. Both types of motion switch are shown in Figure 7-10: the mercury switch on the bottom and the ball switch on the top. Keep in mind that mercury is a toxic substance, so avoid breaking the glass casing that contains the mercury ball.

As shown in Figure 7-11, the pulse shocker schematic is not much different than the last project using the same transformer, but we have added the secondary winding into the works. By applying power to the secondary, we induce a much higher voltage across the primary windings, essentially

reversing the job that this transformer was once doing in the appliance it was taken from. Depending on the ration of windings, you could expect anywhere from 100 to 500 volts on the output, and I promise you that the strength of the jolt will certainly be more intense than the last version of this project. Again, the current is very low, but you will indeed feel the sting of many angry electrons through your joints. As shown in Figure 7-11, the motion switch momentarily completes the circuit so that the primary winding is induced with a large voltage, which is delivered directly to the hands of the unsuspecting user. Also note that often the secondary winding will have several wires, with three being the most common. We are looking for the pair of wires with the least amount of turns, so you can either measure each pair with an ohm meter

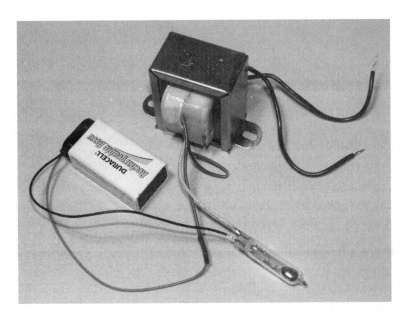

Figure 7-12 *Wiring completed*

to determine the pair with the lowest resistance, or simply look at the color coding of the wires. A three-wire transformer secondary will often have two wires of the same color and another different colored wire. If this is the case, use the different colored wire, and either one of the same colored wires for the highest power output.

Wiring is very easy, as shown in Figure 7-12—just a battery and a switch connected to the transformer. Notice that one of the secondary wires is simply unused, so it can be taped off out of the way. Test the unit on yourself before putting the guts into a case to make sure it's working properly. You should feel a strong single-pulse shock that makes your muscles twitch for a split second as the induction occurs. Hey, I told you it would hurt more than the last unit, didn't I? Remember, if you can't handle the shock from your own evil inventions, then don't expect anyone else to—it's the golden rule of making shock devices, pal. Next, conceal the internal components so that you can see your good buddies jump like pole-vaulting athletes.

The delivery of angry electrons to the victim's hands is similar to the last project, but we are not requiring the connecting of the two wires since this is done automatically by the motion switch. The idea is to ensure that both output wires are touching the operator's hand or hands when the

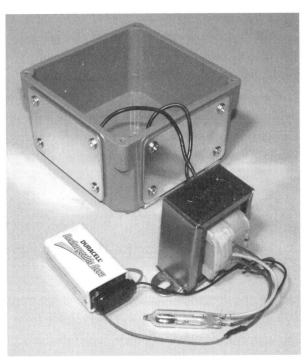

Figure 7-13 *Shock delivery plates*

motion switch is activated so that they get a shock. If they hold one wire per hand, then the shock will travel through both arms and seem a lot more intense than if it was delivered only to one hand through both wires. But make no mistake, they will jump nonetheless. I opted for the shock-to-one-hand option since it was simpler to implement and took up less space. As shown in Figure 7-13, a small plastic PVC box from the hardware store has a pair of metal plates bolted to the sides so that anyone picking up the box will be in contact with both of them. I could have used four plates, but two plates seemed to do the job perfectly. A few alternate ideas may be: a pop can with a few strips of metal used for terminals; a computer mouse with plates on each side; a TV remote; or even a book. The possibilities are endless with this device, and because it can be hidden so easily, even the most suspicious pals can be tricked into picking up the shocker for a good jolt.

Figure 7-14 shows my idea for concealing the pulse shocker in such a way that it is almost impossible not to grab the box to see what it is. I set the mercury switch to about 45 degrees so that the unsuspecting victim of Pandora's box will have it in his or her hand for a few seconds before tilting it far enough to cause the circuit to close. I also padded the mercury switch in a wad of tissue since the shock almost always makes the operator instantly drop the box like a smoldering hot potato. Since this unit works unattended, you don't even have to be in the same room when the prank goes down, but why would you want to miss out on all the fun?

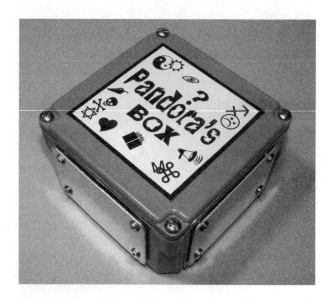

Figure 7-14 *Who can resist the temptation of Pandora's box?*

Yes, curiosity can be a bad thing sometimes, so be careful when you pick up an odd-looking box, especially in the presence of an Evil Genius with a twisted sense of humor. OK, tough guy, if you dare venture into the next project, I will show you how to build a zapper with an output so much stronger that it makes these previous units look tame.

Project 31—Disposable Camera Zapper

Before you dive right into this project, let me warn you that the amount of pain you will receive from this zapper is hundreds of times greater than any of the previous shocking devices. If you thought the last few projects were not effective enough, or if you have one of those buddies who laughs at pain, then this one will deliver the goods, I promise—at 300 euphoric volts, 1000 times per second! Yes, this tiny circuit can send pulses many times stronger than the last shocker at ridiculous rates of 500–1000 times per second, and I guarantee you, there will be no question that you have been zapped! Again, the current levels are very low, but the constant repetition of high-voltage pulses is

what gives this project such a nasty output. OK, if you are not afraid to zing yourself with enough voltage to light up a florescent tube (yes, it can), then let's forge ahead.

As shown in Figure 7-15, all you need is a disposable camera with a built-in flash unit. A camera flash works by charging a high-voltage pulse capacitor through a high-voltage inverter. Once the voltage is high enough, a flasher circuit indicates this to the user via a flashing red LED or small neon lamp. When the user presses the trigger on the camera, all of the capacitor voltage is thrown across a xenon flash tube, which is triggered by yet another stepped up pulse across a

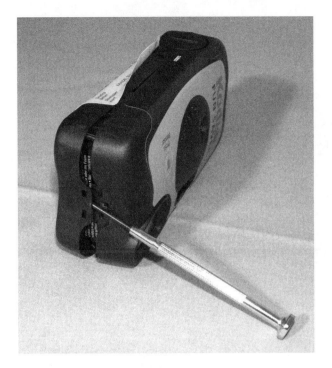

Figure 7-15 *Opening a disposable camera*

Figure 7-16 *Removing the camera casing*

high-voltage transformer. We will eliminate the capacitor and feed the high-voltage charging circuit directly into the zapper's output terminals. Let's begin by snapping the plastic covers away from the lovely innards of our disposable camera. As shown in Figure 7-15, these camera casings just snap apart for easy removal of the tasty innards. Now be careful not to touch any of the wiring, or wear gloves to be safe just in case that capacitor is storing energy. The capacitor can hold many hundreds of volts for weeks at a time, and you will not enjoy an unexpected shock from it.

Crack open the two plastic halves and toss away all of the plastic bits that fall out. These parts are not needed. What we want is the battery (it will still be good even after the camera is used), and the main circuit board. Let me warn you again before you dig right in; do not get your fingers into the leads of the photo flash capacitor, or you will be digging your head out of the ceiling when you fly out of your seat. If you have no clue what I am talking about, remove the circuit board wearing work gloves—trust me. The photo flash capacitor will maintain a charge for a long time, and even months after the last photo was taken, and it may

still have a few hundred angry volts ready to find their way to your nervous system. If you do not know what the photo flash capacitor looks like, read this entire section before you gut the camera. This will be your last warning. The two plastic halves should come away easily if you ply at them with a screwdriver (Figure 7-16), but feel free to get nasty if you need to, it's not like this is an expensive camera.

As shown in Figure 7-17, the circuit board will come free from the rest of the camera as one sparsely populated board with only a few components. It's a truly amazing device considering what it really does in order to charge the capacitor, alert the user and trigger the flash tube. The photo flash capacitor will be the round cylinder about half the size of the AA battery labeled, you guessed it, photo flash. You will want to short the two leads out to drain any leftover high voltage using a screwdriver or some bit of metal. Do not use your favorite screwdriver, and close your eyes when you cross the leads, if there is a charge, there will be sparks, beautiful sparks. The capacitor is able to store a dangerous level of current and voltage, which is more than enough to make the unit hazardous, and we will not be using it in the shocking device, so cut it from the circuit board and add it to your parts box for some other device.

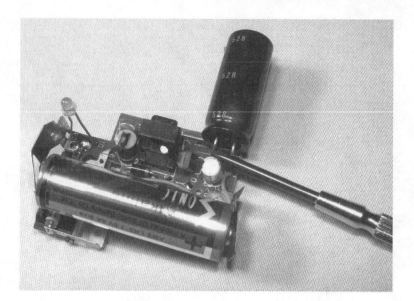

Figure 7-17 *Short that flash capacitor!*

By removing the flash capacitor, we also remove the ability of the camera to store up dangerous levels of voltage and current in trade for an even higher voltage output with no dangerous current to worry about. Our goal is to press a button to deliver an instant shock to the target, and the photo flash capacitor will not allow this because it takes several seconds to charge it up to any serious voltage level. This high voltage is, however, available from the charging circuit, and we do not need the capacitor to harvest the power for our own use. In fact, without the capacitor connected, there is almost three times the voltage available at the points where the capacitor was once connected. OK, if this is really the case then what is the point of the capacitor, you ask? The amount of amperage that the camera flash charging circuit can deliver is miniscule, and although the voltage is plenty high, there just isn't enough power to make the xenon flash tube pop. The capacitor is capable of storing a large voltage and decent amount of amperage, more than enough to set off the flash tube, but it has to collect it slowly from the charging circuit, which is why it takes several seconds to charge between flashes. The charging circuit minus the capacitor can deliver about a thousand volts from the 1.5-volt battery at very low amperage, and

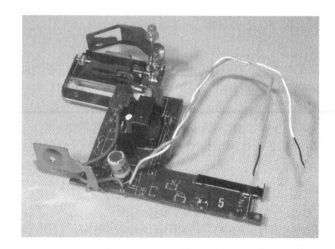

Figure 7-18 *Replace the capacitor with a pair of wires*

although it isn't at any level dangerous to a healthy person, it sure hurts! The output from this zapper will hurt a lot more than if you stuck two metal screwdrivers directly in your AC wall outlet, the difference is there is no dangerous current. Oh, and now that the flash capacitor is not in the circuit anymore, you can take off the gloves. Find a pair of wires and solder them to the points where the photo flash capacitor used to live as shown in Figure 7-18. These will be the output wires that connect to the terminals that make contact with your unsuspecting victim.

You will also need to install a pair of wires for the trigger and the battery, so take note of which one of the small metal clips was the positive side of the battery. The trigger switch is nothing more than a bubble switch that has a small round disk taped over a pair of contacts directly on the circuit board. In Figure 7-19, I installed the wires for both the battery and the trigger switch, and the small trigger-switch disk can be seen at the bottom of the photo. Now you have six wires coming from the camera circuit board—a pair of high-voltage wires coming from the points that once held the photo flash capacitor, a pair of wires for the 1.5-volt battery and a pair of wires

for the trigger switch, which will start the charging circuit. The only important wire to keep in mind is the positive battery wire, as reverse installation of the battery will make the unit fail, and could damage the transistor that drives the oscillator circuit.

You will need to find a 1.5-volt AAA battery holder so you can reinstall the battery as shown in Figure 7-20. If you can only find a dual battery holder, then solder a wire across one of the battery compartments and use that. Hmmm, what about using two batteries you wonder? Well, before you

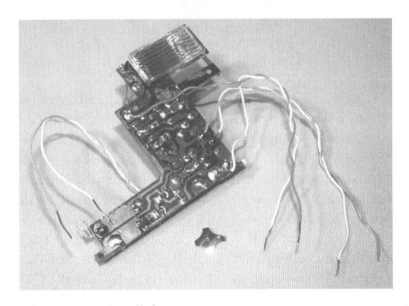

Figure 7-19   *Battery and trigger wires installed*

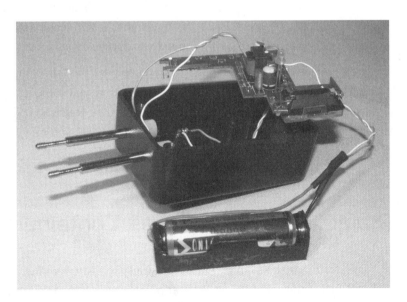

Figure 7-20   *Terminals and trigger installed*

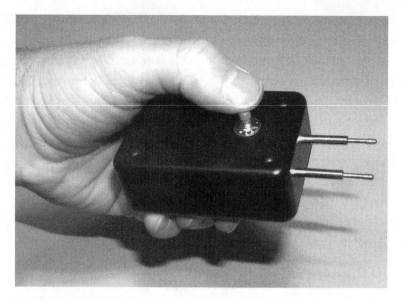

Figure 7-21 *You feelin' lucky, punk?*

start modifying this zapper, try to take a shot with a single battery and then tell me it needs to be twice as powerful—go ahead, what are you waiting for? Yes, indeed the circuit will run with 3 volts or more for a seriously high-voltage output, but I doubt that you will want that after hitting yourself with only a single battery. Also shown in Figure 7-20 is the small box I plan to install the guts into, a pair of bolts used as output terminals, and a push button trigger switch, which is hidden from view under the box. Do not use sharp points as output terminals, a set of small bolts or banana clips will work just fine.

The completed zapper is shown in Figure 7-21, ready to deliver the lightning to anyone foolish enough to get caught between those terminals. Did your pals claim that the shock you gave them from the other projects didn't really hurt (even though their eyes were as wide as saucers)? Well, that won't happen this time, you can be assured of that.

And let's not forget the golden rule—if you can't handle the output, then it's too powerful to be used on anyone else. Want to show off the awesome power of your new toy but can't find anyone brave enough to grab a hold of the output terminals? Try these cool experiments that don't require a human victim. Connect the output to a large fluorescent tube and it will light up. Place the output terminals on some fresh bread and it will start to sizzle and burn after a few seconds. Carefully charge up the original photo flash capacitor by connecting it to the output terminals for a few seconds then drop it across a dime to hear an extremely load crack and a huge spark; it will leave a nice pit where the metal was blown away.

This concludes the "shocking" part of this chapter. Now, I will show you some less painful ways to confuse and scare your good friends for some Evil Genius payback that you might owe them.

Project 32—Hissing Gas Container

Here is a project that is guaranteed to make even the most hardened skeptic wonder if they should laugh or turn and run. This effective little gag

is strikingly realistic looking and sounding, yet 100 percent harmless. The hissing gas container is exactly what its title implies, but instead of any pressurized gas leaking from the container, we will just feed a piezobuzzer with some high-pitched white noise. The result will be what sounds like a spray can letting off its contents. By placing the piezobuzzer inside the container, it will sound extremely convincing. I don't know about you, but if I walked into a room that had a hissing container that looked like it contained some high-pressure gas, I would be heading for the hills without a second look! Let's build it!

You will need a piezo element (refer back to Chapter 2—Truly Annoying Devices for a detailed description on these), a few basic electronic components and a container that looks as though it could contain high-pressure gas or liquid. It will also add to the effect if the piezo element can fit inside the container so that even the closest inspection will make it seem as though the hissing sound was actually coming from the container's opening. The perfect container for this project is a metallic thermos or water bottle, since these look a lot like small high-pressure cylinders, and offer plenty of room to insert the electronics. In a pinch, you can cut away the bottom of the container to install the guts, but if they fit directly into the container, then the final product will look more realistic. Figure 7-22 shows the top of my spun aluminum water bottle with just enough room for the piezo element, 9-volt battery and small circuit board to fit inside.

To make a sound like a hissing gas leak, we will need a source of white noise. White noise (like the sound of an unused section on the FM radio band) is a combination of all possible frequencies at random intensity levels. The sound of ocean waves is a good example of white noise. The circuit shown in Figure 7-23 will generate white noise, which is fed into the piezo element so that only higher frequencies are given off, making a sound very much like a hissing gas leak. The odd-looking circuit shown in Figure 7-23 seems to contain an error, since the collector of transistor Q1 is not connected to anything and the emitter is hooked up backwards. Well, this is no error, and it exploits a little-known fact that the reverse-biased emitter-base

Figure 7-22 *Find an appropriate container to contain the parts*

junction of a common NPN transistor will provide a good source of white noise. This noise signal is then amplified by transistor Q2 and then fed into the LM 386, 1-watt audio amplifier IC to produce a strong output to the piezo element. You could actually use a small speaker instead of the piezo element, but the hissing will not sound as authentic owing to the lower frequencies the speaker will pass.

Go ahead and put the circuit together on your breadboard so you can test component values and audio output. Since we are forcing Q1 to do things that it's not really designed for, some fiddling may be in order to produce the best sounding white noise. I found that common 2N3904 transistors worked well, but feel free to try any type you may have laying around your junk pile. Another thing you may have to mess around with is the values of C4 and R5. If you find the white noise sounds too synthetic, then try some other values, or even add another 2.2M resistor in series with R5. The white noise should sound like a fresh bottle of pop hissing as you unscrew the lid, not like an audio oscillator gone bad. When you do have your white noise hissing brilliantly, reassemble the circuit on a bit of perforated board that is no wider than your 9-volt battery so the entire unit can fit through the opening of the container. Figure 7-24 shows my completed circuit board ready for installation.

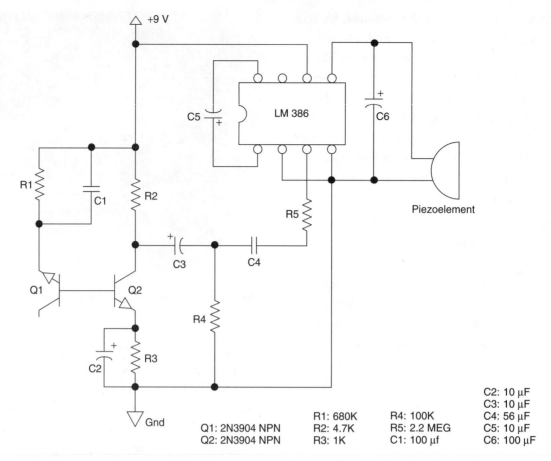

R1: 680K   R4: 100K   C2: 10 μF
R2: 4.7K   R5: 2.2 MEG C3: 10 μF
R3: 1K   C1: 100 μf   C4: 56 μF
                         C5: 10 μF
                         C6: 100 μF

Figure 7-23 *White-noise generator schematic*

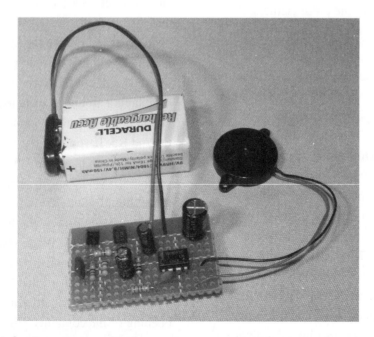

Figure 7-24 *Completed noise-generator circuit*

Figure 7-25 *The last thing you want to see hissing!*

I didn't bother with an on–off switch because it's just as easy to drop in a fully charged battery and let it run for hours of fun.

The piezo element should face the opening of the container so the sound is as loud as possible, and a good way to ensure this is by placing the circuit, battery and piezo element into a tube that is simply inserted into the container. If the container is conductive, then ensure that your circuit board is insulated, or there will be some real toxic fumes leaving that bottle as your transistors fry. Now I'm sure an Evil Genius like yourself needs no assistance in coming up with a label for your hissing container, but let me show you mine

(Figure 7-25). Yes indeed, a hissing container with a biohazard warning on the label. Even those who know your evil ways may think twice about taking a closer look at this baby!

The hissing biohazard container is always a hit at parties, and is a sure method of making an entrance if the bottle comes rolling into a room right before you do. This gag is so easy to set up as well. Just hide the bottle in a quiet room or cupboard and wait for the next person to find it. Glove compartments, sleeping bags, refrigerator, packsacks, all good places to put a surprise!

For our next prank, we will be closer to the front lines, with a hand-held radiation detector.

Project 33—Radiation Detector

The true Evil Genius can turn ordinary junk into amazing works of art designed to deliver a good practical joke with precision. Once in a while you may need a sophisticated-looking bit of machinery to help you pull it off. In this case, our ruse is a fully working radiation detector, or at least so it may seem. This prank works great when the

recipient has no idea that you are the ultimate Evil Genius prankster, since it requires a little bit of setup, possibly done by another friend in on the prank. This fake "fallout" detector is nothing more than a variable resistor-controlled audio oscillator that makes cool sounds as a "radiation level" meter moves around just like the real McCoy, of course,

Figure 7-26   *A large analog meter and enclosure*

it cannot detect anything, so don't try to use it anywhere you might suspect of being radioactive—just run!

To set up this prank to seem convincing, have a friend bring the victim to an area where you are waiting with the device, or show up at the friend's location wearing some official-looking outfit and knock on the door, claiming that there has been a radioactive material spill into the sewage system or something silly like that. Wave the wand over your friend's body and let the machine make a little noise by only turning the hidden knob slightly "Whew, you are safe, luckily." When you get to the victim, crank the knob so the machine screeches like mad while the radiation meter pops over to the danger level, and give your best look of terror as you start asking questions like, "Where have you been?", "What did you touch?" You could also just show up on the scene and detect fallout on people or objects as well for that spur-of-the-moment gag. Feel free to build the machine any way you like. As long as the control knob is hidden from view and easy to operate, you will have loads of fun finding all that radiation at levels that would make a badly maintained nuclear waste dump look clean!

To build this device, you will need a large analog meter (the bigger the better), a cabinet to place the meter and components into, and a few odds and ends to fabricate a crude radiation detector wand. Figure 7-26 shows the large analog

meter I found at a surplus junk store and the enclosure I plan to use. The most effective placement of the meter will be in full view of the fallout victim, so a strap that allows you to wear the detector box around your neck would be ideal.

Also shown in Figure 7-26 is an old coil-type phone handset cord that will look good as a connection cable between the radiation detector handle and the main unit. There really won't be much on the inside of the enclosure and hand-held part, but the more "industrial" everything looks, the more convincing the effect will be, so choose a heavy enclosure and good solid switches and cables. The analog meter can be pulled from an old volt meter, stereo amplifier, battery charger, as well as many old electrical test devices. These meters will normally respond to a few volts in order to set the needle flying across the entire range, so just about any type of meter you find can be adapted to work in this circuit. The first thing you will want to do is snap apart the plastic cover to remove the current indicator plate in order to replace it with something much more ominous looking. "DC amperes," or "Decibels", is not a label that will scare anyone, so I pulled my meter apart as shown in Figure 7-27 in order to convert it into something much more evil.

The plastic front window will probably just pop right off with a little twist of the screwdriver, and the plate with the indicator numbers may be held on with a small screw. Be careful not to bend the

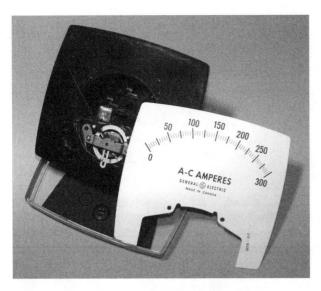

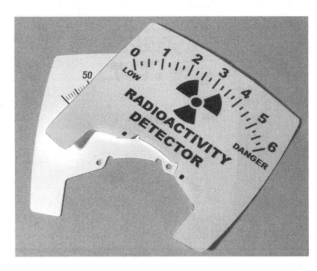

Figure 7-27 *Removing the original meter panel*

Figure 7-28 *Printing out the new meter panel*

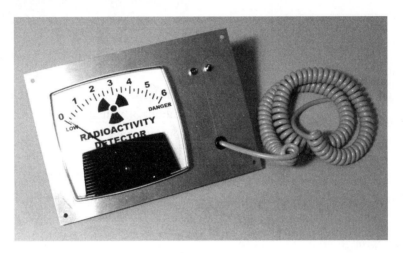

Figure 7-29 *The main unit controls mounted*

delicate needle or damage the return spring when you hack into the meter, just carefully take out the current panel as shown in Figure 7-27 if you can. Some smaller meters may be glued together, so you will have to slip your new printed label into place and then tape it secure. As shown in Figure 7-28, "Radioactivity detector" looks much better than "DC amperes" as it was labeled on the original meter. I simply printed the new label from a graphics program, then traced around the paper using the old panel as a guide. The new panel is then taped right over the old one.

The plastic window is replaced on the meter once the new panel is installed, and now you have a brand new meter that looks like the real deal, no

matter what it happens to be measuring. The next step is to secure the meter to the main unit along with the cord that will connect the radiation detection wand and the on–off switch. Mounting the visible parts was easy since my enclosure came with a removable aluminum top that could be cut and drilled. As shown in Figure 7-29, I cut the hole for the meter body, bolted it in place then added an on–off switch, indicator LED and a hole for the wand cord, which will be connected to the hidden potentiometer that controls the unit. You could get fancy and add a few other gadgets into the wand such as a cool light system, or even another oscillator circuit or buzzer for extra show. This design can take on any form your imagination can dream up. I decided to keep things simple and

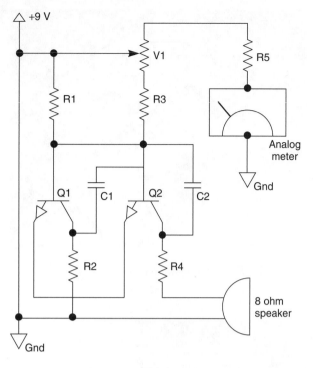

Q1: 2N3904 NPN	R1: 10K	R4: 100 ohm	C1: 0.01 μF
Q2: 2N3904 NPN	R2: 1K	R5: 10K	C2: 1 μF
	R3: 10K	R6: 100K POT	

Figure 7-30 *Audio oscillator and meter drive schematic*

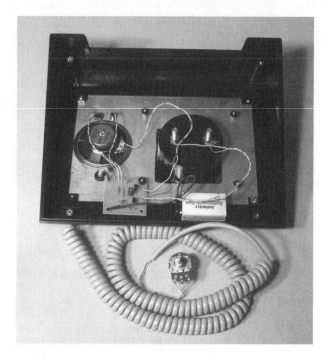

Figure 7-31 *Components mounted*

plain looking, just like a real military-style Geiger counter would look.

Now it's time to dig through your junk box for a pair of transistors, some resistors, capacitors and wires in order to cobble together the circuit that makes this unit come to life. The schematic is shown in Figure 7-30, and as you can see, it is nothing more than a variable audio oscillator that can be set from a low hum to a high-pitched screech by adjusting the 100K variable resistor V1. This variable resistor is actually mounted in the wand so you can conceal the fact that you are manipulating it when you wave the wand around your victim's body. Of course, you could also mount it in the main box if you like, as long is it is not too obvious that this is what makes the meter and sound work. The basic audio oscillator is formed by the two NPN transistors, and the variable resistor not only adjusts the frequency of the audio tone, but it also makes the meter swing

from zero to the maximum. Depending on the amount of voltage it takes to max out your meter, you may have to play around with the values of V1 and R5 in order to get a good balance between how high the audio pitch gets as your meter swings to the full end of the scale. Other than that, feel free to mess around with the component values in order to create some cool effects, or even add more than one oscillator for some really confusing sounds as you move the variable resistor.

Once your audio oscillator and meter are getting along together, move the components on to a perforated board and solder them together for mounting inside the main box. As shown in Figure 7-31, my radiation detector is fully functional and only needs some type of wand to be fabricated in order to conceal the variable resistor at the end of the cable. Remember, bigger and more complex is better when you are trying to baffle your victims with this device, so dig deep into your junk collection for some good parts to make the detector wand from.

I was trying to keep the look of my machine as clean as possible since I planned on giving it a full military-style camouflage paint job, so I just used

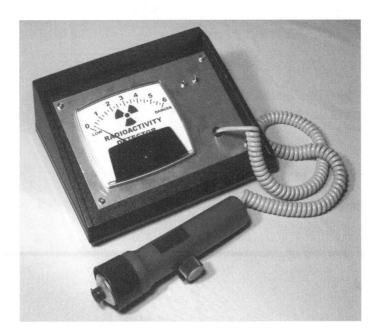

Figure 7-32 *Dude, I think you might glow in the dark!*

an old flashlight for the radiation wand, drilling a hole in the side to install the thumb-operated variable resistor. I also turned the flashlight head around so it looked more like the end of some sensor than a flashlight as shown in Figure 7-32. If you want to go all out, a flashing light show inside the wand would really add to the effect, the possibilities are endless. If you are a freak for details, then do an image search for Geiger counter and see how you can make this project look exactly like the real McCoy for those hard to trick

friends that might know a flashlight from a real Geiger counter tube. Yes, you will be a real hit with your circle of friends when your real-looking detector sends everyone running.

Well that sums up the shock and awe chapter. We have barbequed, shocked, zapped, poisoned and then radiated our friends all in the name of good fun (at their expense mostly). But hey, let's not forget the title of this book, and always remember what comes around usually goes around.

Chapter 8

Machine Hoaxes

Project 34—The Magic Light Bulb

This is a classic hoax often sold in various magic catalogs that you can make yourself inexpensively using some simple components. The finished product will actually be far superior to the fake-looking, plastic units that are often for sale because it will look perfectly authentic, and give off a lot more light due to a high-brightness white LED, rather than an incandescent light bulb. You will need an ordinary screw-type light bulb, a bright white LED, and two small coin batteries that will fit inside the base of the light bulb. The white LED and a pair of 3-volt CR2032 button batteries are shown in Figure 8-1.

These two batteries will provide a total voltage of 6 volts when in series, but the white LED is only rated for 3.6 volts, so what's the deal you ask? Well, since these batteries can only supply a small amount of current, the LED will run just fine, and could probably handle three of those tiny batteries without any problem. The maximum amps (discharge rate) that those batteries can deliver is not nearly as much as the LED can draw, so we find some balance here. If you are not sure if your LED will run on the two- or three-button cells, then just stack them on top of each other and drop the LED leads across the batteries as shown in Figure 8-2 to see what happens. If your batteries cannot deliver enough current, the LED will only glow dimly, and if there is too much current, the LED may get hot in a few seconds or glow with a slight blue hue, indicating that it is getting too much current.

The pair of CR2032 button cells worked perfectly with my 3.6-volt white LED, and I could even run three batteries up to 9 volts, although that did not leave much room inside the bulb casing to install everything. Now for the fun part—trying to get into the base of the light bulb without breaking the bulb or yourself! You will of course, need an ordinary household light bulb, and a pair of work gloves to perform this operation, just to be safe. The plan is to grip the bulb and attempt to force the metal base free from the glass so that both parts are intact. This process may be extremely easy, or very difficult depending on how well the base is glued in place, and you might have to resort to holding the base with a pair of pliers (wrapped around a cloth) if the bulb is really stubborn. The light bulb can be new, used, or even burned out because we are going to generate light with the LED, so the filament's condition is not important here. A well-used bulb may actually be easy to separate since the glue has had time to age and bake from the heat generated by the bulb, but try to avoid a dirty old bulb with a black spot in the center, as it will not pass the light from the LED as well. Figure 8-3 shows my light bulb undergoing a lobotomy as I twist it back and forth with a pair of work gloves for protection against broken glass if the bulb should retaliate.

The bulb should give up its base after a little bit of muscle is applied, so hopefully you won't have to resort to pliers, which could bend the thin metal or crack the glass bulb. When you do get the base

Figure 8-1   *Two small batteries and a white LED*

Figure 8-3   *Do the twist? Lights out Charlie?*

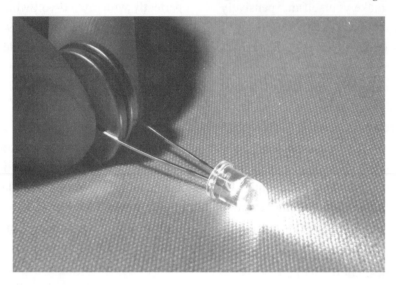

Figure 8-2   *Test the white LED with your batteries*

free, pull or cut the two fine wires that connect the base to the bulb filament so that the tube that leads to the inside of the bulb is unobstructed. This way, the LED can slide into that space. It may look like the bulb vacuum is open now, but in reality, that tube leading into the bulb is still sealed at the end, so the vacuum is still in tact, and no air can escape the bulb. You should now carefully scrape away any leftover glue crust from both the bottom of the glass and the inside of the base so that the two pieces can be easily fit back together. Figure 8-4 shows the base removed from the bulb and the excess glue cleaned up from both parts.

If you have had no luck removing the base from your bulb without breaking something, then try

another bulb until you get a good base and a good bulb from your operation. You could smash a bulb to get a good base, and then rip away the base with wire cutters to get a good bulb. When you do have a good base and a good bulb, you will need to solder a pair of wires into the base so that one wire is connected to the screw part and the other to the point at the bottom of the base, just like the two original wires were once connected in the original bulb. As shown in Figure 8-5, the wires need only be a few inches long, since they will be packed into the base when you are finished.

Figure 8-5 also shows the two batteries held tightly together by a plastic zip tie. This is done because tape will not hold the wire and batteries

Figure 8-4 *Base removed from the bulb*

Figure 8-5 *Wires soldered to the bulb base*

tight enough, and you should never attempt to solder a wire directly to a battery or it may get too hot and explode. The rest of this project is easy. Wrap one of the base wires around the zip tie on one side of the battery pack, solder the other wire to one of the LED leads, then wrap the other LED lead around the zip tie on the other side of the battery. Once that is done,

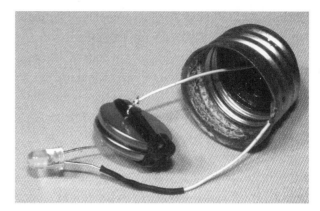

Figure 8-6 *The bulb base becomes a switch*

the LED can be switched on by short-circuiting the bulb base using a ring on your finger or some small metal object hidden in your hand. Remember which LED lead goes to the positive side of the battery when you connect the wires (the long lead).

Figure 8-6 shows the completed LED circuit, ready to be crammed back inside the light base so that it is completely hidden from view. To re-connect the base to the glass bulb, wrap a bit of tape around the glass so that the base has to be forced into place, or use a bit of hot glue to hold it there. Unless you use the magic bulb like a flashlight, the batteries will probably never need to be changed, so don't worry about how well the base is secured to the bulb. The best way to perform this trick is to have the light bulb sitting around somewhere in a room and then ask someone if he or she can hand it to you. The light bulb looks and feels like an ordinary bulb, so there will be no suspicion as it is passed along to you. Now, use your metal ring, a very fine wire, or some other small hidden metal object in your hand to make a connection between the base of the bulb and the point at the bottom to close the circuit and light up the bulb. Look surprised, hold out the bulb and let everyone see the mysterious self-powering light bulb glow like magic in your hand. As your baffled friends request the bulb, they will be fully confused at the fact that there is nothing out of the ordinary about it, and it certainly does not glow when they handle it. Show them that there is nothing in your hand, then take the bulb back, making it light back up again. The trick will fool a person as long as you can keep a straight face.

As shown in Figure 8-7, this completely normal-looking light bulb sucks energy from the fourth dimension, and never needs a light socket. The thin metal wire I have along my finger is making the connection at the base to complete the circuit, but cannot be seen by anyone, even at close range, so this trick works flawlessly. The only thing you should remember is that this bulb should not be left lying around when you are done baffling your friends, or someone may try to install it into a light socket! After all, it does look like a regular bulb, but I can assure you, that if you put 120 volts AC into that battery pack and LED, there will be smoke, lots of smoke and a nice spark, so hide the bulb when you are done with it. Our next machine hoax will also seem to draw energy out of thin air, so keep on reading if you like to confuse your scientist friends who should know better.

Figure 8-7 *Look ma, no power!*

Project 35—Coin-minting Machine

Before you consider building this project, it is a good idea to check your local laws regarding the duplication of money; in some places it may be unlawful to mint your own coins from metal slugs using an old CD-ROM drive. Of course, we are not really going to mint any coins, but we will make it look as though this is happening by the use of a sneaky internal "switcheroo" and some mechanical noise. The idea to make a machine look as though it is minting coins came to me when I was looking at an online magic catalog and saw a dollar-bill printing machine. This device is nothing more than a roller that feeds in a bit of money-sized blank paper from the top and rolls out a dollar bill from the bottom, and it is so simple to build and figure out that I did not bother trying to make one. In true Evil Genius fashion, I came up with an elaborate machine that includes an autoloading door, mechanical noise maker and a hidden magnetic coin switcher, just to print a measly quarter! Of course, if

you want to fool a person, you need to baffle them with a lot of bells and whistles in order to conceal what is really going on—it is the key to a successful hoax. To build the coin-minting machine, you will need an old CD-ROM drive, a few metal slugs about the same size as the coin you plan to mint, and a bit of thin wood or plastic. The only part of the CD-ROM that needs to function is the tray, since it will be used to load the slug and then output a real coin. Figure 8-8 shows the sacrificial lamb along with a few metal slugs and a quarter.

A suitable metal slug about the size of a quarter can be found by prying the disk that plugs one of the wiring holes on a common electrical box. The slug is slightly larger than a quarter, but you can tell your onlookers that the edges will be trimmed in your new coin minting machine, so it needs to be slightly larger than a coin. The slightly larger slug will also help to hide the illusion when you are loading the machine as you will soon see.

Figure 8-8 *A few metal slugs, a coin and a CD-ROM drive*

Figure 8-9 *Parts for your junk bin*

Let's start by removing everything out of that old CD-ROM drive until nothing remains except for the loading tray and the motor that makes it open and close. Start by removing every screw you can see to split the metal cabinet open, and then continue taking out all of the screws until the parts are strewn all over your workbench. You will not need the main board, laser head, spindle motor, or any of the parts that are not directly involved in opening and closing the drive tray. Most DVD and CD drives will contain three small DC motors: one for the tray-loading mechanism (the one we are keeping intact), one for turning the spindle, and one to move the laser head

up and down the track while reading the data. Since our machine needs to make some fake coin-printing sounds, the laser head motor will be perfect for this because it is a standard DC motor with only two connecting wires. Figure 8-9 shows all of the leftover parts after reducing the drive down to nothing except for the tray-loading mechanics.

The loading mechanism will consist of a plastic tray riding along a pair of sliders, one of which is cut into a gear rack so that the loading motor can drive a small pinion gear along the gear teeth to open or close the tray. The motor will be the size of a pop bottle lid and will be connected to a belt-drive

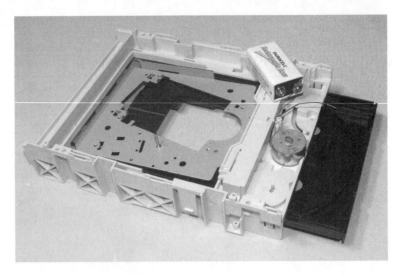

Figure 8-10 *Testing the loading mechanism*

Figure 8-11 *Drive tray insert with slug hole in center*

reduction system to slow down the speed of the gear that runs along the gear rack. If you cut the connector off of the end of the motor wire pair and hook it directly to a 9-volt battery, the tray will either open or close depending on the polarity of the battery. Normally, this motor runs on 5 volts, so the tray will open and close very quickly when connected to a 9-volt battery, but this is OK, and will not harm the motor. Figure 8-10 shows the stripped-down loading mechanism being tested by connecting a 9-volt battery directly to the motor wires.

The loading tray will now be modified to carry the metal slug into the unit, so a bit of thin wood,

cardboard or plastic must be cut so that it can be bolted to the tray where the CD or DVD used to sit. The thickness of this insert should be no more than the thickness of two quarters, or it may get stuck between the drive cabinet when the door is closed. I found that the thin wood backing found behind dressers and cheap stereo speakers was the perfect size for this job as it was about the same thickness as heavy box cardboard. A hole slightly larger than the metal slug should be drilled through the center of the inset as shown in Figure 8-11 (the slug is inserted into hole). A piece of thin cardboard or sheet metal will need to be placed under one side of the hole as well so that the quarter and slug do not fall right through the insert hole. I used a bit of cardboard.

The insert needs to be about the same thickness as two quarters because it will carry an actual quarter with the slug placed over the quarter when it is being loaded. The quarter is hidden from view by the slug until the drive tray is fully closed at which time a small magnet will remove only the slug so that the quarter will be the only thing left in the insert hole when the tray is opened. The operator will press the "coin-minting" button after the tray is loaded and closed to make a mechanical noise to add to the effect, but the actual switcheroo will happen instantly inside the machine without any noise at all. The drive tray insert will be bolted directly to the drive tray using some small nuts and bolts, or machine screws as shown in Figure 8-12, but ensure that the bolts do not hit the surrounding mechanical parts as the tray is opened or closed. This should be tested with the drive casing installed over the loading mechanism to ensure that there are no clearance issues.

Once you have installed your coin and slug tray insert and checked it for proper operation using the 9-volt battery, you can now create the magic switcheroo device, which is nothing more than a small magnet placed over the insert hole inside the CD-ROM cabinet. Most CD drives will have a small round metal lid in the top cover placed exactly over the center of the motor spindle, so this is where we will mount our lifting magnet. This small metal lid is perfect for our device because it will allow easy removal of the hidden slug after we baffle our friends with our new high-tech counterfeiting operation. Most likely, the metal lid will be under the manufacturer's sticker and need a bit of prying to remove the glue seal that holds it in place. If for some reason your drive casing does not have a removable lid, as shown in Figure 8-13, then you will have to drill or cut one yourself. The opening needs to be as large as the slug so you can remove it once the magnet has done its job.

The small ceramic magnet shown in Figure 8-13 was taken from a fridge ornament and is thin enough to avoid getting in the way of the drive tray once the lid is placed back over the drive cover opening. This magnet does not need to be very strong in order to do its job, and it is better that it is not, since you only want to lift the slug away

Figure 8-12 *Drive tray insert installed*

Figure 8-13 *Magnet glued to the round lid*

Figure 8-14 *Magnet lifts the slug not the quarter*

from the quarter, not both at once. As shown in Figure 8-14, the small magnet will easily lift the slug from a distance of about half an inch, but it does not have enough power to lift the quarter at the same time. The magnet is held in place with a bit of hot glue and then the lid is placed back over the drive casing for another clearance test. If all goes well, your slug and quarter will go in, and only the quarter will remain once the tray is opened; this is the real magic behind this hoax.

The tray motor will need to be connected to a few switches in order to reverse the polarity of the battery for loading and unloading the machine. The easiest way to do this is by connecting the battery through a double-pole, double-throw switch, or a pair of single-pole double-pole switches as shown in the schematic Figure 8-15. As you can see the two switches labeled "direction switches" simply reverse the polarity of both battery lines so that the motor can be made to open or close the tray. I did

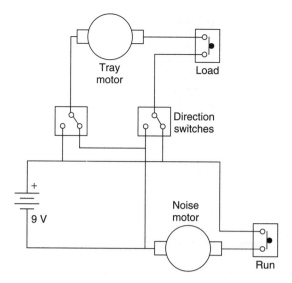

Figure 8-15 *Wiring diagram for tray and noise motor*

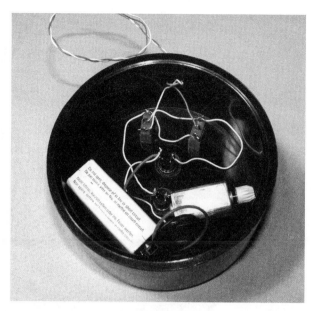

Figure 8-16 *Switches and noise motor inside a PVC cap*

not have a double-pole, double-throw switch handy, which is why I ended up with a pair of double-throw, single-pole switches instead. The operation using the two switches is exactly the same but you have to engage both in the same direction to reverse the polarity. The other switch connected to the tray motor is a momentary push button to activate the motor once the direction has been set. Simply set the direction switches to open or close, then press the momentary switch to make the tray drawer open or close. The other switch labeled "run" activates the noise-making motor, which makes some grinding sounds much like an old dot-matrix printer running. Tell your onlookers that the slug is being cut by some high-tech laser mechanism inside the box.

Figure 8-16 shows the completed wiring of the operation switches placed inside a PVC tubing cap. This cap will be held over the slug-loading lid with a bit of Velcro so the slug can be easily removed after the coin has been minted. Notice that the noise-making motor is also hidden inside the PVC cap, and it is positioned so that the included bevel gear can rub directly against the inside of the cap, making a familiar cutting and grinding sound when the "run" switch is pressed. The two wires that exit the cap will connect directly to the CD-ROM tray loading motor through a small hole in the cabinet.

With the switch box mounted over the round lid in the drive case, your coin-minting machine is ready for use. Next time your buddy needs to borrow a quarter for the phone, or when you need some parking change, bring out the machine and surprise them with your elaborate plan to never run out of spare change. The best way to set up this hoax is to have the real quarter and a slug between your fingers so that the quarter cannot be seen. Now, hand your buddy another slug to look at while you place the hidden quarter and slug into the device, keeping the real quarter out of sight. Now that the quarter and slug are safely loaded, explain that your new machine will cut a quarter out of the slug using a high-power ruby laser mounted to a series of X and Y plotter motors (make up something good). Now, move the direction switches to the close position and press the tray button to bring the slug and quarter into the device. The slug will now be picked up by the small magnet, and you can press the noise-making motor for a few seconds while you explain how the laser is etching the surface of the slug into a nice new quarter. By this time, your buddy may be quite impressed by all of the noise and technical jargon that you have spewed while the quarter is being cut by the machine. When you are ready, move the direction switches to the open position

Figure 8-17   *Three steps to never-ending spare change*

and press the tray button to reveal a nice new quarter, freshly minted by your fantastic invention! Figure 8-17 shows the machine at work.

The machine is very convincing, and will baffle your friends every time you can't seem to find needed spare change. If you want to make the machine look less like a CD-ROM drive, then replace the front plastic tray cover with something of your own design, and enclose the drive and operating switches into another cabinet with more lights, bells and whistles. With a little mechanical imagination, the unit could be made to pull the switcheroo on other objects as well such as printed money, credit cards, even food!

Project 36—See Through Walls

This device is so simple to build that you won't even have to warm up your soldering iron, but the end result is a hoax that will fool just about anyone, even those technical skeptics who should know better. This machine will convince an onlooker that you are looking directly through a wall and can be used indoors or outdoors with equally good results. This hoax requires nothing more than a portable video player that can be loaded with your pre-made clip of the device scanning a wall, then the space behind the wall.

The video player is placed in a box that looks like some kind of bizarre video camera, and then you simply copy the motion that you made while you were recording the original video. This may sound simple, but I can tell you that this hoax is one of the most convincing hoaxes I have ever contrived, and it fools everyone the first few times they see it, even those who know such a device could never exist. Just stand in front of your house one day and let the neighbors look at the "view screen" of your device in pure amazement as the image changes from the front of your house to your living room as if it were looking right through the wall. The effect is even more convincing due to the fact that the image is moving back and forth as you scan your house from left to right in sync with the playback image. The instant that the video changes from the outside of the house to the inside can be made more realistic by inserting some cool video effect on your computer editing program if you have one, and you can have as many clips on the video player as there is room for, allowing you to look through multiple walls, houses or buildings. The two key components you will need are a portable video player and some type of lens from a security camera, or optical device as shown in Figure 8-18.

Figure 8-18 *A portable video player and a lens*

Figure 8-19 *Mounting the lens to the enclosure*

The lens can be any size and shape you like, and it does not have to perform any other function besides looking like a lens, so use your imagination for a parts' source. Any video player will work for this hoax, and no modifications or damage will occur to the device, since we are simply going to mount it in a cabinet so that it looks like part of the covert spy machine. You will need a cabinet that will contain the video player in such a way that the view screen is easily visible and so that you can get access to whatever buttons you need to press in order to power up the player and get the video clip to play. On my device, the play button will power

up the device and play the first video clip, so I only needed a single hole in the cabinet in front of this button. Once you have found an enclosure that will contain the video player, find a way to attach the fake lens to the front of the box as shown in Figure 8-19. The lens should be at the opposite side of the cabinet to where the view screen will be, just like any video-recording device.

Once the lens is mounted, insert the video player into the cabinet so you can mark the holes that need to be cut. Place a piece of paper placed over the video player and trace around the video screen and control buttons to make a template that can be used to mark the holes in the cabinet. This way you will avoid scratching the video player. If you want to cut a nice square hole, you can use a handy notching tool that takes small rectangular nibbles from plastic or thin metal; these can be purchased at most hardware stores, and make this job a lot easier than using a jigsaw of file. Figure 8-20 shows the video screen and play button holes cut out of the enclosure so that I can just drop my video player inside anytime I want to do this hoax.

Since I wanted the ability to easily remove the video player when the hoax was over, I just placed the player into the cabinet and stuffed it with toilet paper so it would hold the player in the correct place

Figure 8-21   *Video player mounted in a non-permanent fashion*

Figure 8-20   *Enclosure holes cut for the video player*

without the need to fasten anything in a permanent fashion. If the device really could see through walls, then I would have no problem sacrificing the video player to the cause, but alas, I have yet to invent such a device! As shown in Figure 8-21, a wad of toilet paper is bunched up and stuffed into the box to hold the video player against the side of the cabinet where the cutouts have been made. The view screen and play button holes line up perfectly with the video player, and there are no signs of the original video player hidden inside the device.

OK, now let's make the video that will convince everyone that you are really looking through the walls with your invention. Take your video camera and choose a target wall to look through. If you want to baffle the neighbors, use the outside of your house, and for indoor hoaxing, pick a wall in your living room that has a bedroom on the other side. Stand in the place you plan to perform the hoax and record 5 seconds of still video and then make two or three slow and steady passes from left to right along the wall while keeping the video camera about eye level and parallel to the ground (this motion will be easy to duplicate). Now, stand on the other side of the wall and do the same thing, trying to keep the motion as fluid and steady as you can. Once you have these two video clips, download them to your computer editing program and bring them into the timeline as shown in Figure 8-22 so you can crop and merge the clips together. If you have never used the simple editing software that came with your video camera, then get out the manual and go for it. It really is easy.

The goal is to mix the two video segments together so it looks as though the camera that made them had the ability to focus from the wall you are looking at instantly to the room on the other side. Start by trimming the first video clip of the 5 seconds of still video and outside wall footage so that there are no jumps at the start and

so that it ends on the last pass along the wall before you hit pause to move to the other side of the wall. Now, clean up the second clip the same way and drop it on the timeline so there is a nice clean fade between the two clips and what appears to be the same left to right motion. When you play this back, it will seem as though the camera was never moved and that the wall simply faded away to reveal the room on the other side. If you are good with your editor, try a lightning bolt, or some caustic melting as a transition to make it look like the wall was digitally faded out of the video. No matter how you do it, just make sure that the left-to-right panning is smooth through the entire video clip before you save it. Save the mixed video, then use the conversion utility that came with your video player to import it into the device. You are now ready to convince your buddies, neighbors

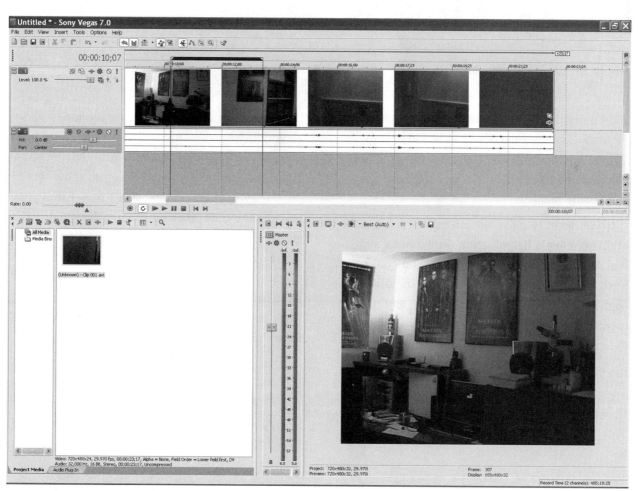

Figure 8-22 *Mixing the two video clips together*

Figure 8-23 *Privacy is a thing of the past with this device*

and family that you have the ultimate spy device. Figure 8-23 shows the device in action as it looks through my hallway wall (left photo) right into my computer lab (right photo) in an extremely convincing manner.

If you want to make the effect look even more realistic, have a helper get in on the hoax and include them in the video as well. You could have a partner run into the room on the other side of the wall you plan to look through, and then record them waving their hand or something on the edited video. When they come back out of the room, they can say "Did you see me? I was waving my hand." This will make the entire hoax seem so realistic that you may find the "men in black" knocking at your door if you show this to the wrong people. Oh, and don't forget that your helper should be wearing exactly the same clothes on when you shot the original footage. There you have it, the ultimate invasion of privacy machine that will fool anyone, and make them think you can watch them through their walls any time you like. The neighborhood will never be safe again!

Chapter 9

Mind Benders

Project 37—Rigged Lie Detector

A lie detector (also called a "polygraph"), is a sensitive electronic device that measures electrodermal activity (skin resistance), respiration rate and blood pressure. Rumor has it that an expert can read the output from a polygraph and determine if the subject is telling the truth by comparing the data with a set of test questions given to the subject. There is much debate whether these devices are accurate, and research by some experts in psychology indicates that it only takes minimal effort to control your body responses and fool the device. Of course, none of this debate matters when using the lie detector presented here, since it is the operator who actually determines the outcome. Ironically, this device uses the same principles as a lie detector. It uses the operator's skin resistance to trigger a hidden touch switch so that the truth or lie indicators can be controlled covertly. The schematic for this extremely simple device is shown in Figure 9-1, and at first glance it may seem like there are a few components missing. The two tabs labeled "lie on" and "true on" are simply conductors that, when touched, will create a voltage and capacitance change on the input pins of the LM358 operational amplifier IC causing the appropriate LED to light up.

These two touch-switch terminals can be made from any conductive material, which makes concealment very easy. In my design, I connected the touch-switch wires to the small metal screws that hold the plastic enclosure together so they are completely hidden and can be used without

detection by simply sliding a finger over the appropriate screw hole. You could paint over the conductor, since there will still be enough potential difference to trigger the op amp as long as your finger is directly over the touch-switch area. The LM385 op amp is a good choice for many projects because it has a wide operating voltage range, only needs a single supply, and is a common and inexpensive part. Practically any LED can be directly connected to the op amps outputs without requiring any current-limiting resistor, or you could add a switching transistor to drive a larger load, such as a buzzer or relay, to make the results of your device a little more dramatic. There are so few parts that it's almost not worth using a circuit board, but to keep things neat and clean, I added the op amp and dual resistors to the small bit of perforated board as shown in Figure 9-2.

Also shown in Figure 9-2 are the two screws that fasten the plastic box lid being used as touch-switch points for covert operation. For the indicator LEDs, a bright red LED for "Lie" and a green LED for "True" were used. They were sufficiently bright, even though the op amp can only source a minimal amount of current. The actual probes that your fibbing friend must grab on to can be made from anything you can find in your scrap pile since there is no electrical connection to the circuit board. You are secretly in control of the device, so the lie detector probes can take any form you like, from an elaborate set of probes connected to the subject's body, to a fanciful-looking space-age

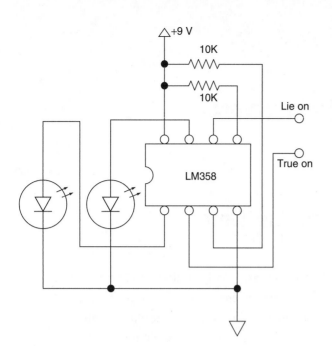

Figure 9-1   *The lie detector schematic*

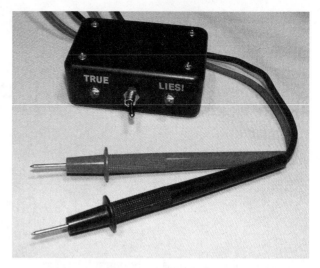

Figure 9-3   *You won't fool this lie detector*

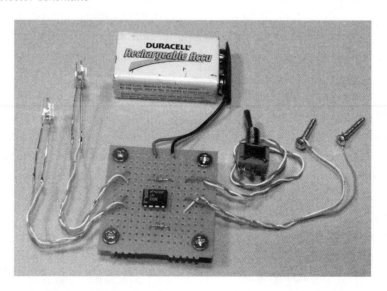

Figure 9-2   *Lie detector circuit assembled and ready*

helmet that reads brain waves. I opted for a set of probes cut from a defunct multimeter, and simply inserted the wires through holes in the plastic cabinet then tied a knot to keep them from being pulled out. To operate the device, turn the power switch on, have your deceitful buddy grab on to the probes, then secretly slide your finger on to one of the touch-switch contacts to control the outcome of the lie detector. If you want to set up the prank to make it seem more realistic, run through a few test questions so you can prove to the subject and bystanders that the device is really working. "Are you human?" "Are you wearing a blue shirt?" Then when it comes time to prove that the subject is telling lies, the effect will seem more convincing since the unit did seem to have the ability to distinguish between obvious truths and lies during the test questions. Figure 9-3 shows my simple lie detector with the multimeter probes and indicator LEDs. The two rear screws are connected to the touch-switch wires so I can slide my fingers over the appropriate one while out of view.

With this lie detector in hand, no amount of psychological training can be used by the subject to fool the device, since the operator is in total control.

Now you can reveal the truth (your version of it) in any situation, regardless of what lies may be spewing from your subject's mouth!

Project 38—The Dog Talker

This fun device can be used for a prank in which you need a simple way to transmit your voice or some recorded message to a remote location. I call it the Dog Talker because it was initially designed to allow our pal, DJ Dogster, to communicate with us humans. The little device which is worn around his collar is simply labeled "K9 Translator" in big bold lettering in order to grab the attention of anyone interacting with our K9 buddy. Once they start to check out the box, I can either talk directly through the radio receiver hidden inside using a microphone, or play some prerecorded message from files saved on the computer. The device can also be used in "stealth mode" without the obvious label so that it is hanging under his collar and not very noticeable. This way, DJ simply strolls up to one of his human buddies and begins to talk. If we have company over, I can set my computer to play some funny wave files through the device at preset times so it seems as though DJ himself were doing the talking. "He, he, he, he … Scooby Snacks"—you get the idea. Of course, this easy-to-make transmitter and receiver combo can be used to project sound for many other funny pranks, and because it has an audio input, any audio device with a line output, such as a computer, radio or MP3 player, can be used as the source audio. You will need a set of inexpensive kid's "walkie talkies" in order to build this project. Practically any set will do, and they all work on the low-power 27-MHz or 49-MHz license-free band, so you won't have to worry about the "radio cops" busting down your door if you mess with them. Figure 9-4 shows the $5 walkie-talkie set that I found at a bargain shop for hacking.

If you plan to make the Dog Talker or need to hide the receiver in a small area, then look for the smallest walkie-talkie set you can find so it can be easily concealed. The units I hacked are a bit on the bulky side, and I have seen inexpensive sets a quarter the size, but since DJ would have no problem carrying the 2 × 5-inch box on his collar, these units will do just fine. In this project, one of the units will become a permanent receiver, reduced to the smallest possible footprint, and the other unit will become a permanent transmitter, always transmitting so that the receiver never hisses when there is no signal. The transmitter will also be modified so you can plug in a better quality microphone or connect it directly to some audio source for playing back prerecorded or timed messages. Dig right into one of the walkie talkies, removing the plastic cover so that you have only the circuit board, battery and speaker remaining. If the speaker is glued to the cabinet like mine was, then unsolder the wires first, so you can pry out the speaker without having the circuit board in the way to avoid damaging any of the components. There may also be an electret microphone included in the cabinet that will also need to be pried loose, so do the same thing with it. In Figure 9-5, I reduced the completely functional receiver to the circuit board and speaker.

If you want more than a 20- or 30-foot range, then you may need to install an antenna back on the receiver, but you may be surprised at how well a 2-inch piece of wire will work when compared to the original antenna that was included in the plastic casing. My unit actually had the same range without any antenna connected to the receiver as

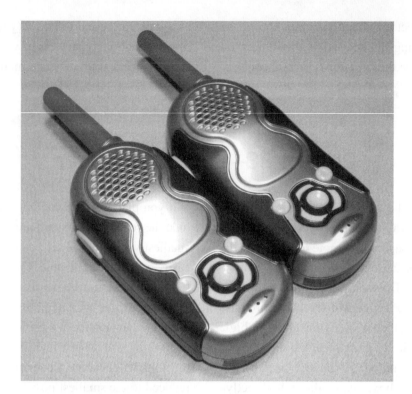

Figure 9-4   *A set of kid's walkie talkies*

Figure 9-5   *Removing the plastic cabinet*

long as the transmitter was placed in a good spot without any metal object between it and the receiver. With your receiver unit gutted, try to pack everything into the smallest possible container you can find so the unit will be easy to conceal and lightweight. The largest parts of the receiver may

be the battery and speaker, so keep that in mind when searching for a plastic box to contain the unit. If you find the speaker a bit too wide for the cabinet, it may be a good idea to install a smaller unit, but you will lose a bit of volume in the process. With a little searching, I found a plastic box that fit

Figure 9-6 *Re-fitting the receiver in a smaller box*

the circuit board, battery and speaker perfectly, and it is shown going together in Figure 9-6. The original plan was to sew the guts into a doggie vest, but this little plastic box would make the Dog Talker more fun to use because people could notice it from a distance and come to investigate out of pure curiosity.

Now that the receiver is ready to go, let's modify the transmitter so it is always transmitting and will never hiss due to signal loss. We will also solder an external input jack in place of the internal microphone so your voice can be projected to the receiver with much better clarity and this will also allow external sound sources to be used. A ⅛-inch stereo jack, like the type you find on the end of a headphone, will be the best type of jack to install due to its common use in audio equipment and small size. It is very easy to find a microphone or dubbing cable with this end so you can connect directly to your computer's sound card or an MP3 player. There are two types of walkie talkies: those that include an electret microphone and those that use the speaker as a microphone when the transmit button is held in. It does not matter which type of unit you have, since you will remove the microphone or speaker, and run the two wires to the external input jack. Try to find a space on the original plastic casing where you can drill a hole to install the jack. This may be a bit of a challenge on

the smaller units, and if there is no free room inside the case, you can run the wires through a small hole and solder the jack outside the case. I managed to find a tiny space for the jack between the battery compartment and then unsoldered the wires from the unit's built-in microphone and soldered them to the new jack. Now another microphone or external audio source could be plugged into the walkie talkie, making it an all-purpose audio transmitter. One annoying factor was that the transmit switch would have to be held down in order to transmit, so I decided to trace the contacts with my multimeter and solder them in the transmitting position so the unit would always be sending the signal to the receiver. This constant transmission will also stop the receiver from hissing when there is no signal, a dead giveaway that a transmitter is being used. Figure 9-7 shows the new ⅛-inch audio jack installed in place of the walkie talkie's built in microphone.

If you cannot figure out which pins to short in order to fix the transmit switch so that it is always engaged, you can force it closed with some hot glue or shrink wrap, and then remove the plastic knob so it no longer sticks out of the casing. If you have no microphone for testing the new input jack, use the original microphone that came with the unit by soldering it to a male ⅛-inch jack. You can also feed the output from your computer or

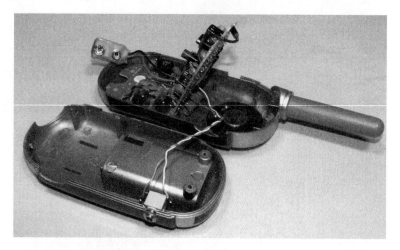

Figure 9-7 *External input jack installed*

Figure 9-8 *Ready to transmit any audio signal*

MP3 player with the volume down low into the transmitter to see how the audio sounds on the receiver unit. When everything is working, put the transmitter back together, and you are ready to use your new broadcasting station. Any time you need to send an audio signal a few hundred feet, this system will be ready to perform, and it can be used like a low-power pirate radio broadcaster to send your tunes to the garage, or even as a simple room bug to monitor a remote location. The transmitter is connected to the output of my computer's sound card in Figure 9-8, sending some funny "dog talk" to DJ's collar as he roams about in the front yard, talking at surprised passers by.

If you want to play the talking dog prank, then there are several options. Hiding the receiver under a collar or sewing it into some doggie clothing will

give the most surprise, especially if you are hiding out of view and watching the action. You could watch through a window as someone you know comes up to pet your K9 friend, and then bark "Hello!" through the transmitter, or prepare some funny prerecorded message to playback from your computer or MP3 player. Placing the box out in plain view and calling it a translator is also fun, since it will make some people curious and encourage them to come over and talk to your dog. "Whatcha got there pal?" they might enquire as they bend over to pet your buddy. "Woof! What's up, dawg," your canine pal responds. Yes, it can be quite entertaining to interact with curious humans who are surprised that your canine talks! You can also have the transmitter hidden in your pocket, connected to an MP3 player or some sound

Figure 9-9 *Woof! I make this look good!*

playback device so you can have your dog talk to people as you meet them while out on a walk. There are a few voice output ICs on the market that can spit out a few hundred English words by simply sending the correct serial data, so you could have a world of fun talking to humans through your pet. To demonstrate the "K9 Translator," my good buddy DJ Dogster sports the receiver on his collar before heading outside to communicate with the humans (Figure 9-9).

Well, I hope you have fun with this project, and find many uses for this simple audio transmitter. It's always a hoot when things start to talk or sounds come from places that they normally wouldn't.

Project 39—Telepathy Transmitter

Now you can have the same power that talkshow psychics, divine healers and those memory gurus have—the power of a hidden radio transmitter! Yes, the "magic" they have is usually hidden in a shoe or inside a tiny earpiece so they can be fed instructions, or data about the person whom they are trying to convince that their psychic abilities are genuine. Our transmitter will not take the form of an earpiece audio transmitter, which would be too obvious and expensive to make for home-grown magic. Instead, we will create a system that can be completely hidden under clothing that will allow the sender to press a button and transmit a mechanical tap to the receiver. The person with the transmitter will be told a secret number or color, and then telepathically send his or her thoughts to the person with the receiver hidden on them. Now, you might think that it would be difficult to communicate with a simple system like this, but for simple mind-reading games where the receiver only has to guess colors or numbers, it is perfectly suited. Many casino cheaters use a system similar to this, sometimes hidden in their shoe, and they have no problem transmitting numbers and card names in a hurry, so it can be done. Morse code can also be used with this system, if you are up to the challenge of learning it, although you could still just tap out the number representing the letter you want to send (1-26) for short words.

This project can take on many forms, depending on where you plan to show off your telepathic abilities and what type of clothing you plan to wear at the time. For impromptu indoor use, a simple belt system worn by the hoaxster is easy to use and will allow the sender to covertly tap a hidden switch under a short. For outdoor use, a shoe-mounted tap switch might be better and would certainly be very difficult to detect, just like the would-be casino cheaters use, but don't get any bright ideas pal, the house always wins! I decided to use the guts from an inexpensive mini RC car to make this unit, owing to the fact that all the electronics are already made and the system is very small. I needed to reduce the transmitter to the smallest package possible and then modify the little car electronics to move some mechanical device that would allow the receiver to feel the taps against his or her body. Figure 9-10 shows the little RC car and transmitter before hacking surgery.

Remove the car's cover to get access to the small receiver circuit board. Be careful not to rip any wires off the board when you remove the casing, since you will need to know where the battery, motor and antenna wires connect in order to connect them again. The two wires that lead to the main drive motor are the ones you will need.

The transistor that turns on the motor will have enough capability to switch a tiny solenoid, relay coil or another motor used as the mechanical activator that taps the person carrying the receiver. In my unit, the power wires were red and black, and the motor wires were white and blue, so the color coding was obvious, but don't assume this to always be the case or you will end up frying some electronics. The motor that powered this RC car was so tiny that I decided not to use it as a mechanical activator because soldering wires to it would probably destroy it, and it probably would not have enough torque to make a noticeable tap on the person wearing the receiver. Figure 9-11 shows the disassembled RC car, a new motor I decided to use as an activator and the tiny rechargeable battery that was inside the car.

I could have used the tiny battery that was included in the car, but then I would have to recharge the receiver whenever I wanted to use the unit, which would be annoying and not allow the receiver to operate for long periods of time. Since this tiny battery was 1.2 volts, I decided to replace it with a single 1.5-volt AA battery that would run the unit for days at a time, allowing the prank to be set up hours before showing off our new telepathy magic. The antenna wire can also be cut down to an inch or two, and should still allow enough range to have the transmitter and receiver in separate rooms to truly baffle your audience.

Figure 9-10 *A micro-sized RC car and transmitter*

Figure 9-11 *Tiny RC car autopsy completed*

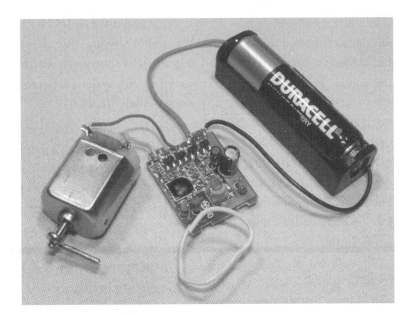

Figure 9-12   *New battery and activator installed*

For an activator, any electromechanical device that can be felt if pressed against your waist or leg will do the trick, and this could be a tiny motor with a counter weight, pager vibrator motor, relay with the cover removed, or even a home-brew coil with a magnet floating around the center. All you have to make sure of is that there is enough current coming from the two wires that are used to run the RC cars motor to power the device you plan to use. The activator should also be silent and easily canceled to avoid detection when doing your telepathy act. Figure 9-12 shows the new battery and activator made from another toy car motor with a bit of copper wire soldered to the shaft to make an arm that would strike the receiver's body.

I could now switch on the RC transmitter and make the activator arm swing around and strike the table as I engaged the "drive" button on the remote control. There wasn't a great deal of force driving the activator, but it was easy to feel the strikes when placed on a belt and worn around the waist or leg, so it would work perfectly. I could have made the receiver a lot smaller by finding a way to use the original tiny motor and rechargeable battery, but since this unit will be worn under loose clothing, its current size should not be a problem.

The receiver circuit board was wrapped in a bit of electrical tape and then placed inside a small round container along with the motor and activator arm. The battery was installed outside the container so the unit could be shut down when not in use by removing the battery. Both the battery and container were then tie-wrapped to a belt that could be worn on the body of the telepathy receiver so that the activators arm would strike them when the remote control was engaged. Figure 9-13 shows the completed receiver belt unit ready to fool everyone the next time we are invited on national TV to show off our amazing telepathy and mind powers.

I was originally planning to gut the RC transmitter and install it on another belt, but since

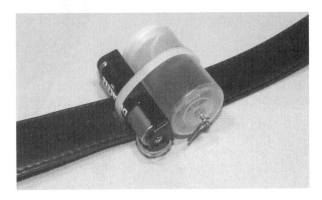

Figure 9-13   *Telepathy receiver belt ready for use*

it was already installed in a nice small package, I decided to simply solder a microswitch to the two terminals that would activate the receiver so that I could place the transmitter in a pocket and then run the microswitch to a convenient place on my body. The microswitch could be run along the user's leg and placed in a shoe, or strapped behind the knee so that leg flexing would activate it, or it could even be replaced by a tilt switch ad placed just about anywhere. The receiver was so small it could also be strapped to the body under some clothing to avoid detection, so this wired microswitch system would work out very well. The microswitch should be as sensitive as possible and not make an obvious click when activated. You can pull these tiny switches from many small appliances, such as printers and telephones, or you could use your imagination to create your own pressure-sensitive switch. A nice flat switch that can be worn in the shoe might be made from two copper pennies insulated from each other at the ends so that stepping down a bit might close the circuit, or you might try making some type of switch that can be activated by flexing a leg muscle inconspicuously as you transmit your secrets telepathically to the receiver. The only goal is to make the switch easy to

operate and difficult to detect, that way you can efficiently tap out your messages to the receiver. Figure 9-14 shows my very basic remotely wired touch switch system using the original remote-control body intact.

Now you can rehearse your telepathy magic with a partner to become efficient at sending messages with the telegraph-like device you have just created. If you know Morse code, then the possibilities are endless, and you can engage in a conversation in very little time by tapping out the words while the receiver speaks them out to the amazement of everyone watching. For simple communications, have a bystander write a series of numbers on a page then hand it to you for telepathic transmission. Simply tap out the numbers in sequence with a long pause between them so the receiver can pencil them down. You could also code colors using numbers, or simply count out the position of alphabetical letters to transmit simple words such as "cat," which would become three taps, pause, one tap, pause, and then ten taps for each letter. There are many ways to prove your telepathic abilities, or cheat at games (did I say that?), so I will leave you to your own evil devices on this one. Have fun, and remember to mention my name when you are on national TV.

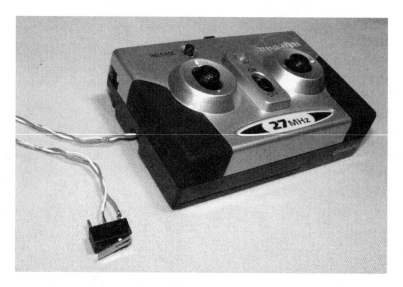

Figure 9-14 *Remote-control transmitter and microswitch*

I do not believe there are any "holy grails" of mind control, maybe except for the CIA's "MK-Ultra" project. But, seriously folks, there must be something to subliminal mind control (also known as "advertising"), because billions of dollars are spent bombarding our senses with product logos, slogans, patterns and familiar scents into our brains. However, some "subliminal" advertising is blatantly obvious, playing on our most basic instincts, and some is so well hidden that it is usually only discovered by accident. Do a Google search and type in the phrase "subliminal advertising pop cans" to see one of the most famous cases of subliminal advertising to see what I mean. Subliminal mind control is not only limited to visual objects, it can also take the form of hidden audio tracks conveying verbal messages, hypnotic alpha-beat frequencies embedded in audio, ultrafast words displayed between frames in video, and even certain smells purposely vented outside an establishment to draw people in. Yes, these tactics really work, and I will show you how you can experiment with subliminal messages encoded in both audio and video clips by using your computer and some basic editing software. If this experiment really does work for you, then imagine how much fun it will be to encode the message "wear a yellow raincoat to school today" on your buddy's MP3 player.

Let's begin with embedding a hidden voice message into an audio file so that your every wish can be fed directly to the minds of your subject through an MP3 player. In addition to the basic mixing software, you will also need a copy of the music they typically listen to, access to their player or music directory, and some way to record your voice into the computer, such as a multimedia microphone. For a good description on getting a microphone plugged into your computer, refer back to Chapter 5, Project 18. I use a very popular sound-editing program called Sony Sound Forge, but for this simple editing any program that will allow you to record and mix sound files will work. Start by recording your spoken commands so that the recording level is smooth without clipping, distortion or background noise. Speak your commands as if you were giving instructions to a hypnotized person, using slow and steady tones, making each word as clear as possible. Keep your instructions simple and short, such as "you will pay back the money your borrowed to Jack" or "it is time to upgrade the computer." You will want to repeat the message multiple times during the song to implant the message, but keep it so subtle that the listener is unaware of the message. Figure 9-15 shows the sound mixing software open with two audio clips, one of the songs I plan to implant the message into (bottom clip) and the spoken message itself (top clip).

Notice that the sound sample does not saturate the top and bottom of the sample window, which would indicate distortion and clipping. The spoken sample appears to be at the same volume level as the song sample, which indicates a good recording. Once you are satisfied with your spoken audio clip, select all of it or part of it by highlighting the wave file in the window, then right click and select "Copy." Almost all audio editing software has this basic function, and you can also select "Copy" from the toolbar if it is not available on the right mouse click. This will copy the highlighted portion of the source audio file to the clipboard so it can be pasted on to a new or existing audio track or clip. Now select the place in the destination audio clip by clicking the mouse button in the clip window. This will become the place that the subliminal message will be mixed in, so try to choose a place that has a lot of sound, not the intro or end of the music piece. With the audio position marker clicked into position, you will be able to right click and either select "Mix" or "Paste" depending on your editing tools. For Sony Sound Forge, the best choice is "Mix" as shown in

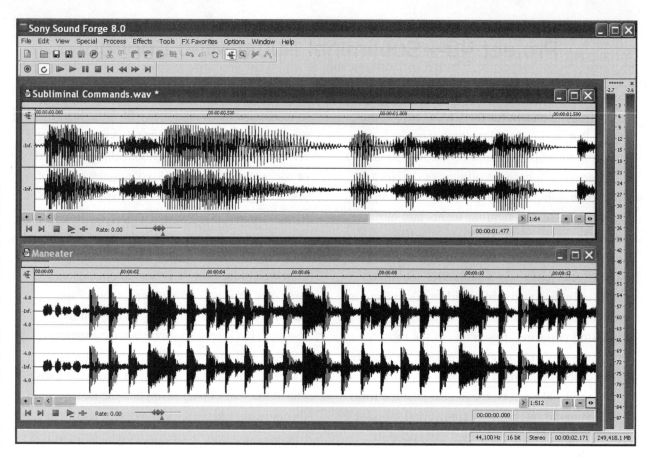

Figure 9-15 *Song and subliminal message open for editing*

Figure 9-16, as it allows you to choose a source and destination volume as the clips are mixed together.

Remember that the subliminal message should be very difficult to hear, even if you know its there, so choose a low mixing level, such as 10 or 15 percent, and test the playback. You should barely be able to hear the voice commands that you inserted, but they should also be subtle enough that, if you weren't listening for them, they would be undetectable. Now you have a true subliminal message. Depending on your goals, you may want to insert more messages throughout the song to ensure that you have total delivery of the "package" over the course of the song. Just make sure that you don't paste the subliminal message over any dead spots in the song, or your ruse will be discovered. It's a good idea to play back the entire song to hear if the subliminal message is well blended before you save the song for import back into your buddy's MP3 player. When you are happy with the

mix, choose "Save As" from the file menu and see if your editing software can directly encode an MP3 file. As shown in Figure 9-17, Sound Forge allows MP3 files to be saved at 96 Kbps, a perfect format for direct upload to the MP3 player without having to use the player's import software tools. Don't forget to save the original song so you can revert it back the way it was once you have delivered your subliminal message to the subject and had all your evil wishes completely fulfilled.

Now all you have to do is wait a few days to see if your subliminal message takes hold as your unsuspecting subject gets his or her groove on. If your message is perfectly mixed and the commands aren't too outlandish, you just never know what might happen. Of course, if you are commanding your buddy to buy you a new entertainment system and then run down the street wearing nothing but a tutu and a pair of flippers, then you might not get what you want! Doesn't hurt to try, though.

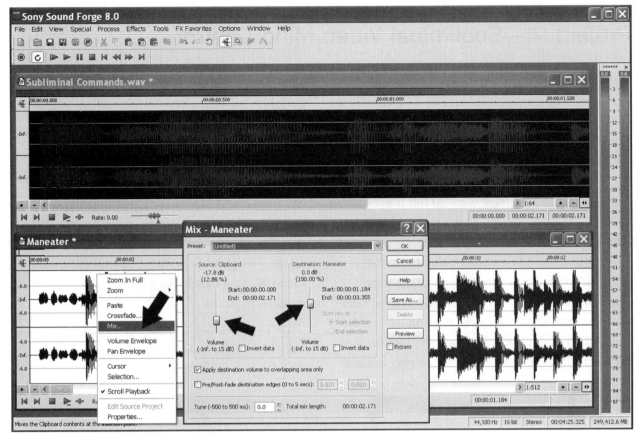

Figure 9-16   *Mixing the subliminal message with a music file*

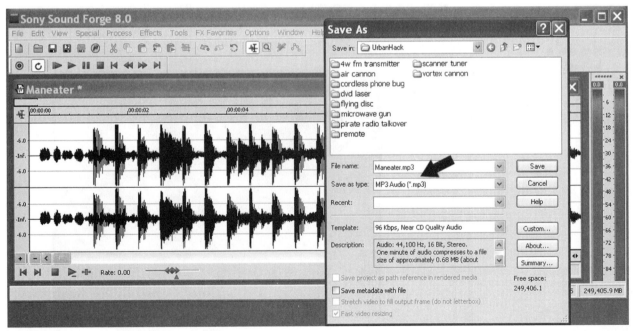

Figure 9-17   *Outputting the newly mixed song*

Embedding very short text messages or images into movies or television broadcasts is an effective method used in advertising. Decades ago, advertisers tried inserting a single frame of subliminal video in a television broadcast so that the mind had only an extremely short glimpse of the message because there were 30 full frames of video drawn every second on a TV set. At such a short duration, the message was not even noticeable, even if you were looking for it, but some "scientists" had determined that the mind could actually absorb the message directly into the subconscious. Of course, these types of manipulative things are not done any more (so they say), but I thought it might be cool to play around

with them because it is not difficult to insert these ultrashort messages into downloaded movies or burned DVD videos. We will use basic video editing software, which is often included with digital cameras and video appliances. The video editing software needs to have basic cut and paste functions (most of them do), and some type of simple text or title generator so you can insert your millisecond message into the movie or video clip. Again, I use a product from Sony called Vegas Video because they make good software, which is reasonably priced.

Start by choosing a video clip or movie that you will show to the hoax victim. You might want to rip a DVD to your computer and then recode it

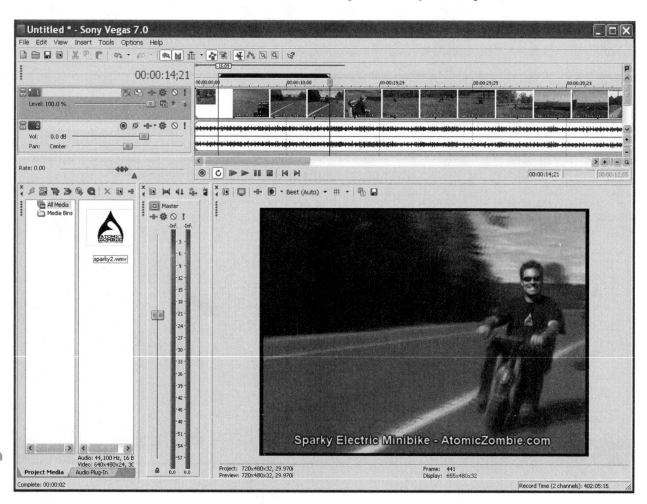

Figure 9-18   *Importing a video clip for editing*

back to disk, or simply alter an existing video clip. Whichever method you choose, the first step is importing it into whatever video editing software you plan to use. As shown in Figure 9-18, I imported a short video clip on to the timeline of a crazy dude riding an undersized electric mini-bike at dangerously high speeds.

Now you must choose a position in the video clip where you would like to insert the subliminal message. This should be somewhere in the movie or clip where you know the observer will be watching intently. In a movie, there may be several places you would like to insert the message for optimal effect, such as suspense scenes or any other place in the clip you know will have the viewer's attention. In the timeline, click on a position you would like to insert the message and then use the razor, split or cut tool to break the clip

into two segments at that point. In Sony Vegas, the tool is called "Split" and is found on the "Edit" toolbar as shown in Figure 9-19. This tool will break the video clip at the exact position of the cursor on the timeline. Almost all video editing software has this basic function, although it may be called something different.

Once the splitting or razor tool has sliced the clip into two parts, click the rightmost segment and slide it to the right so you can open up some space to insert some text-generated media (your subliminal message). Nearly all video editing software has some ability to insert text or titles, and again it may be called something other than "text media" as it is in Sony Vegas. As shown in Figure 9-20, I chose text media from the "Insert" toolbar and then typed a one-line message using a big font. The message "order a pizza now" was

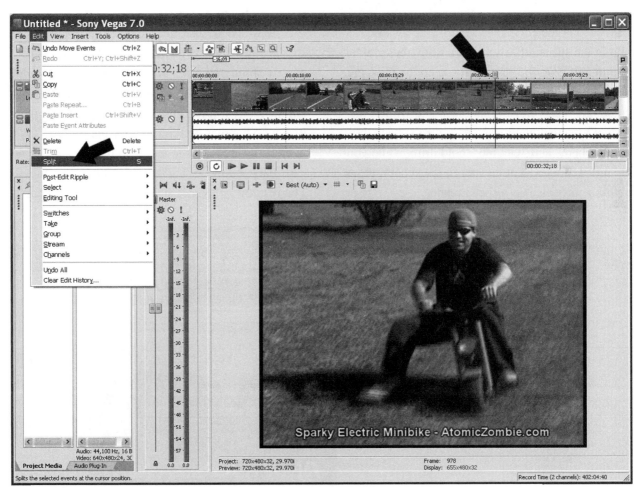

Figure 9-19 *Splitting the video clip to insert a message*

Figure 9-20 *Choose a short message in bold type*

generated at the cursor on the timeline as a short clip in the space left over after cutting the original clip in half. For optimal results, keep the message very simple, use big block text, and make sure the colors are high contrast like black and white. At $\frac{1}{30}$th of a second, the message "Give me back the comfy chair," and "Get me a cold drink while you are up" will likely not register in the subconscious unless your viewer is a speed reader.

Now we need to shrink the text clip to the smallest possible segment, then sew up the entire clip by sliding all three parts back together without any gaps. As shown in Figure 9-21, I made the text clip last only a fraction of a second by shrinking its length handles on the timeline, and then I moved all three clips together so there would be no gaps between the text and video parts of the file. The resulting playback is an extremely small burst

Figure 9-21 *Shortened message now inserted*

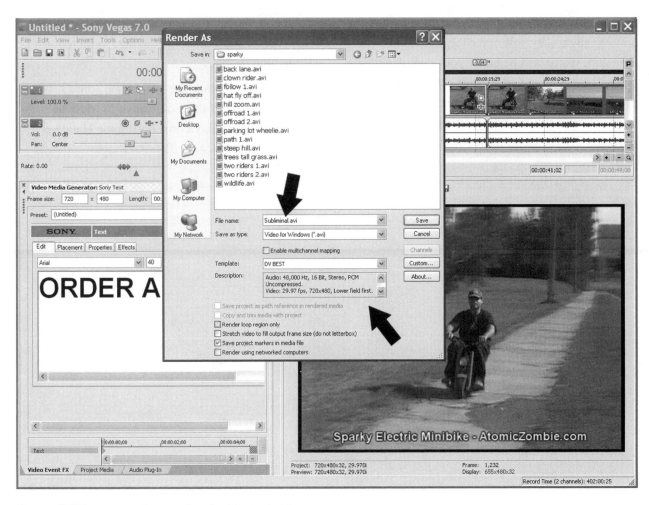

Figure 9-22 *Saving the completed subliminal video*

of text that is almost unnoticeable if you are not expecting it. There will be a tiny gap in the sound as well because the text message has no sound track, but at such a short duration it will be practically unnoticeable during playback. If your subliminal message is easily seen, then it is too long, so keep shortening it until you can barely notice it. To the unsuspecting viewer, it may only appear as a slight glitch in the video file, if they notice anything at all.

Once you are happy with the playback and have inserted your message in one or more places in the video file, you will need to re-encode it back to your hard drive for playback or burn to a DVD disk. There will usually be an export dialog located on the file or edit menu where you can choose to export your newly edited video clip in one of many compression formats. In Sony Vegas, I chose

"Render As" from the File menu and then entered a filename and type as shown in Figure 9-22. For DVD quality playback, it is best to choose a quality compressor with very little loss, such as DV or AVI, so that your DVD re-authoring software can make the best decisions based on a clean file. If you plan to upload your file to the web or send it by email, then a higher compression, such as WMV or MPEG, would probably be better. To get the best results, you will have to do a little experimentation if you are new to video editing software, and if you are really brave, refer to the software manufacturer's manual. After you save your video file, take a quick look at the section that should have the subliminal message just to make sure it is there, although it may only look like a frame glitch unless you can play the file back in slow motion with your media viewer.

Now you are ready to have your friends over for a movie night they will never forget! Before long, the entire world will be under your complete control, carrying out all of your commands with precision! OK, that may be far-fetched, but don't underestimate subliminal advertising until you have had the chance to try it out. A lot of companies spend megabucks to experiment with this type of subliminal messaging, so there must be something to this.

In the next section, we will create some scary goblins and gouls that come alive to frighten unsuspecting hoax victims during the Halloween season, or all year round!

Chapter 10

Halloween Horrors

Project 42—Flying Vampire Bat

You know what irks me? Friendly-looking spooks and smiling ghouls. When did Halloween, "Night of the Living Dead," become so watered down? But, when you come to our house, you'd better be prepared to jump because the place is rigged like a haunted Fort Knox, and you won't find one skeleton with a smiley face anywhere! What you will find are talking mirrors, lurching zombies, bloodthirsty Jack-O-Lanterns and flying vampire bats like the one presented next.

As we all know, bats are creatures of the night that fly around looking for prey, striking without warning to leave victims lifeless and fully exsanguinated. OK, that's actually a load of bat guano, but it sure sounds good on the one night that we get to scare people without consequence. This project can be used to "reel" a small object back and forth across your yard to make it seem as if something is flying towards a person, so it's perfectly suited for giving the vampire bat flight while would-be trick or treaters invade your property looking to gobble up all of your goodies. Let's start by fabricating the towing system that can be used to make little beasties fly across your yard towards visitors. You will need a clothesline pulley and one of those hanging trolleys as shown in Figure 10-1. The trolley is only needed for one of the small pulleys, so if you can find one by itself, that will also work.

Also shown in Figure 10-1 is a spool of heavy fishing wire, which, when strung across your yard at night, will be completely invisible yet capable of

holding up a decent amount of weight. The trolley is only needed for one of the smaller pulleys. Drill or hammer out one of the pins that hold them into the frame to extract the pulley. Now, find a bit of threaded rod or a bolt with a length of about 8 inches that will fit in your drill chuck. I used some ⅜ threaded rod, and I had no problem sliding it into the chuck on my hand drill, but you should check this ahead of time, since the motive power for this project comes from the drill, which must connect to this threaded rod or bolt. Once you have found an appropriate length and size of threaded rod, drill a hole through the small pulley that you took from the trolley so the threaded rod can fit through it to form an axle as shown in Figure 10-2.

Once the threaded rod is inserted through the hole in the pulley, a bolt on either side will lock it firmly to the axle so that you can use it as a drive pulley by grasping one end of the threaded rod in your drill chuck. The two bolts holding the pulley in place are shown in Figure 10-2, but do not crank them down just yet, since the axle will need to be adjusted and placed through the support blocks that will be made next. The support blocks shown in Figure 10-3 are made by cutting a few pieces of 2×4 wood, tall enough to hold the drive pulley in place so that it does not rub on the surface the blocks are placed on. In other words, the drive pulley should be easy to turn when held in place by the two support blocks once you drill the hole in each block for the axle. The washers are placed between the bolts that lock

Figure 10-1 *A large clothesline pulley and trolley*

Figure 10-2 *Bolts hold the pulley and trolley in place*

Figure 10-3 *Axle support blocks made from pieces of a 2 × 4 wood*

their position. Keep in mind that the two support blocks should hold the axle in place without causing any friction, so do not sandwich the pulley too tight with the support blocks. Two long woodscrews placed through the underside of each support block will hold them to the 2 × 4, but you might want to pre-drill the screw holes with a small drill bit first to keep the support blocks from cracking as you tighten up the woodscrews.

You can now connect your hand drill to the drive axle as shown in Figure 10-5. I added a back plate made from some scrap plywood to secure the drill in place while it is being used to drive the pulley, but that is optional. The drill should also be variable speed and reversible or you will only have a bat that can fly one way across your yard at high speed! The support plate for the rear of the drill is nice to have as well because it allows for a one-handed operation, rather than having to hold on to the drill handle at all times when driving the pulley. If you do plan to support the drill on the 2 × 4, make sure the drill body is in alignment with the axel and has a bit of freedom to move so you do not stress the drive axle or the drill chuck.

A little-known fact is that your 120-volt AC-operated drill will also run on a DC voltage as low as 12 volts. If you would like to keep the speed and torque of your pulley to a minimum, just connect the power cord from your drill to a large

the pulley to the axle and the support block to keep the bolt from rubbing against the wood. The axle hole should also be drilled a little larger that the diameter of the axle (threaded rod) so it can turn freely in the hole with little friction.

Both axle support blocks are held to a length of 2 × 4 as shown in Figure 10-4, so that one end of the 2 × 4 has enough room for a drill to sit. The drill side of the axel should also have at least 2 inches of axle sticking out past the support block so it can be securely held in the drill chuck while it is driven. Tighten the two nuts on either side of the drive pulley, add the washers, and then place the two support blocks on the long 2 × 4 to mark out

Figure 10-4   *Pulley and support blocks mounted to another 2 × 4*

Figure 10-5   *Supporting the drill body for one-handed operation*

12-volt battery, such as a security or car battery, and it will turn with 10 times less speed and force. With the drive side of this project ready for action, you need to create the idler side, which will carry the weight of the flying object and wire across your yard. The idler pulley shown in Figure 10-6 was designed to do this exact job, and can easily be adapted for mounting to any pole or wall by placing it on a board. The idler pulley mounting board has a hole drilled in each corner so a rope or bungee cord with hooks can be used to wrap it tightly to a tree or pole across your yard.

You could probably drag practically any creature across your yard as long as the cable you install on the pulleys can take the weight of the object.

Heavy fishing wire can easily support a few pounds of weight safely, which would allow the creation of any number of Halloween creatures such as ghosts, bats, or caped creatures. I decided to create a lightweight black furry vampire bat with glowing red eyes, which would allow me to use very thin fishing wire that would become invisible as soon as the sun started to set. I made the bat by bending a coathanger into the wing shapes shown in Figure 10-7 so they could be covered in black cloth, then fastened to a small plastic box which will also be covered in fur. The small black box will also carry a 1.5-volt battery to light the two small red LEDs to give the bloodthirsty vampire bat a little bit more of an evil look.

Figure 10-6   *Large idler pulley bolted to a board*

Figure 10-8   *Bat wings and glowing eyes added to the plastic box*

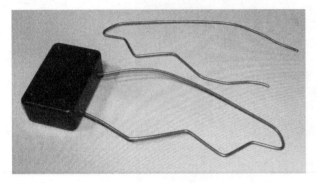

Figure 10-7   *Forming wings by bending coathanger wire*

Figure 10-9   *Ready for flight and ready to attack!*

The black stretchy material was first glued to one side of the coathanger wire using a hot glue gun and then it was stretched around the other side and glued to form the wing. Black paper or tissue paper will also work, and you could simply lay the wire wing down on the paper to trace the area to be cut using a marker. Secure the wings to the small plastic box by placing each end of the coathanger through a hole drilled in the side of the box and then bend the end of the wire so it cannot be pulled out of the hole. The completed wings are shown in Figure 10-8 fastened to the black box along with the two red LEDs that will pierce the night with an evil red glow.

The completed vampire bat is shown in Figure 10-9, ready to fly across the yard towards visitors entering the yard in search of Halloween treats. The bat may not look very realistic close up, but it does put on a good show when it appears to fly across the yard all by itself with eyes glowing red. At a distance of several feet, you might

actually think the bat was the real McCoy, well maybe not with the glowing red eyes.

Installing this project is best done in the light before the unsuspecting trick or treaters head out so you can make sure it functions properly and get used to setting the proper drill speed. Start by fastening the drill to the drive pulley system and then secure it to a table or chair in a place that allows a clean cable run to the target pulley across your yard. Cables need to be out of reach of people and high enough so the bat will seem to come in from above as if dive bombing the victim. Your drive system should also be in a place that does not look obvious yet allows a clear view of the yard and target area so you can know when to launch the bat. You could also place the idler pulley at the target destination and run the drive system from the other side of the yard for a more covert and quiet operation if you have another person handing out the treats. My drive system was strapped to the

Figure 10-10 *Securing the drive system to a workbench*

Figure 10-11 *Evil bat waiting for victims*

workbench in my garage as shown in Figure 10-10, where I can wait on Halloween night to hand out tricks rather than treats. When you are installing the fishing wire, make two or three wraps around the drive pulley to start so there will be sufficient friction to keep the wire from slipping. From there, you will just unspool as much wire as you need to make it to the idler pulley and back.

The idler pulley is strapped to a tree or nailed to a post across the yard so that the bat will travel from above to the target destination. The wire does not have to make multiple wraps around the idler pulley, since it is not required for drive, so just place the wire around it once like it would be

done on a clothesline. The bat or flying critter can be fastened to the wire using a screw or hook placed on the top of its back, which will allow a point to tie the wire. Now, you can test your flying creature by running the drill in both directions to make a back and forth trip across the yard. Start by using a slow speed on your drill, and slowly work up to the desired speed just to make sure there is no excessive friction or anything rubbing on the wire that may cause a breakage. Figure 10-11 shows my evil vampire bat ready to lurch from the top of the tall fence directly across the yard as trick or treaters head to the door expecting to see more friendly ghosts and smiling ghouls. Won't they be shocked when the bat with glowing red eyes does a fast flyby as they approach the front door where many other evil machines await their arrival? Yes, indeed, this is one house in the neighborhood where adults as well as kids can sometimes find themselves jumping for cover—but that's what Halloween is all about.

Project 43—The Haunted Ghost Mirror

This haunted ghost mirror is another fun Halloween prop that allows you to interact with those who are brave enough to venture up to your door after dealing an onslaught of evil trickery in the front yard. At first glance, the mirror looks like a typical black mirror that a practitioner of dark magic might use for "scrying" (watching someone to control or influence them), and since there is no reflection in the dark glass, the trick or treater might ask what the mirror is for. At this point, the mirror seems to come to life as an evil face fades into view as if manifesting from the very darkness. To further surprise the now-captivated mirror gazer, the face in the mirror begins to have a conversation with them, just like the fabled magic "mirror, mirror, on the wall."

For this project, you will need a mirror with at viewing area of approximately 12 inches squared, since there will be a video screen placed behind it eventually. The mirror itself will be taken out and replaced with a plate of clear glass or plastic, so you may simply want to make your own mirror frame instead of taking one apart. A large border is also nice to have, so you can paint it up with symbols to look like a real authentic tool of dark magic. Figure 10-12 shows the rather plain-looking square mirror I decided to convert into my haunted ghost mirror.

The ghostly image will be projected through the black mirror by cranking up the brightness and contrast on an old portable video monitor or security monitor so that only the highlights appear through the dark cloth behind the glass. For this reason, your mirror glass should be about the same size as the video screen you plan to use or just

Figure 10-12 *A 12 × 12-inch mirror with a large wooden frame*

slightly smaller so there is no visible border. As shown in Figure 10-13, I used an old 14-inch color monitor with a composite input so that my video camera could display the image directly. Also shown in the photo is a small black and white security camera, which also works perfectly for this project since image quality and color is not important. Any video screen that can display the image live from your camera will work, and this could be a security monitor, television set with a line input or RF modulator, or even an LCD screen, which would be very easy to conceal.

If your video camera has any special effects such as negative or mirror image, then you can spice up the image on the screen to make a truly bizarre display. Connect your camera to the video screen and then increase the contrast and brightness until you have the brightest possible image with a lot of definition between the light and dark spots.

Figure 10-13 *A monitor that will display a live camera image*

Figure 10-14 *Scary faces from out of the darkness*

Figure 10-15 *Securing the material over the clear glass front*

Since the mirror fabric will only pass the highly contrasted image, detail or even focus is not important here. A little black and white makeup can make you appear very creepy when viewed through the highly contrasted mirror material, so you have a lot of room to experiment, making your face look as evil as possible. The material that you will use to cover the video screen should allow the highly contrasted image to show through, but fade off to pitch black very quickly as the light or brightness of the image fades, this way you can sit in front of a bright light and fade in and out of view like a ghost. A bit of black stretchy spandex seems to work very well when stretched over the video screen, as shown in Figure 10-14. In this photo, I am sitting on a chair with a 600-watt floodlight on my face for maximum brightness, and when I back up a little bit, it seems like my face is fading into the darkness as the image completely disappears. Some of the other materials that gave cool effects when placed over the video screen were: dark nylons, white paper, florescent lighting lenses, and many other thin semitransparent materials.

When you find a material that gives you an effect you like, remove the original mirror from the frame and replace it with a clear plate of glass or a bit of plastic. The material you plan to use will then be placed behind the glass as shown in Figure 10-15, using whatever method works best for you. I stapled the spandex material around the frame to give it a bit of stretch, allowing a bit more light from the video screen to pass through it. Now, your mirror can be painted to look more like an official tool of witchcraft and then placed in front of your video screen on a table top for your Halloween visitors.

The video screen is simply covered up with a bit of dark cloth and then surrounded by various objects, such as candles or other interesting objects that will keep attention focused into the center of the screen. If you like to go all-out on Halloween, then a false wall that will allow the ghost mirror to be hung over a video screen opening will make the effect look very realistic, fooling many who encounter the mirror. A simple false wall can be made from a

Figure 10-16 *I am the ghost in the mirror!*

4 foot by 8 foot piece of wall paneling, or even painted or wallpapered drywall to make it look like part of your house. The video screen can sit on top of a ladder or chair behind the false wall so that the image is projected through a hole in the wall which is covered up by the ghost mirror. You, the actor, can sit anywhere out of view that will allow you to hear what visitors are saying into the mirror, so you can respond appropriately. If you want to make it seem like the mirror is talking, just run the audio output from your camera to a pair of computer speakers hiding near the ghost mirror, setting the volume loud enough for visitors to hear your voice without any feedback. Figure 10-16 shows the most basic ghost mirror setup, using a dark towel to conceal the video monitor behind the mirror as some twisted individual hams it up in front of the camera.

On Halloween night, the ghost mirror will be placed on a false wall and my voice will be lowered to sound like a demon by running the audio output from the video camera through a pitch-change effect normally used for guitar. The special effects on the video camera will make my image negative for a more eerie-looking ghost face, and I will also add a strobe light for additional effects. The ghost mirror and wall will be placed at the front of the house where trick or treaters are greeted, and I will be sitting around the corner so I can hear what they are saying, allowing for a fully interactive conversation to happen between bewildered guests and the ghost face in the mirror. Once again, the only interesting Halloween display on the entire block will belong to the Evil Genius, bwa ha ha ha!

Project 44—Living Brain in a Jar

Have you ever seen some of those twisted gore flicks that air in the wee morning hours, where some crazed mad scientist attempts to re-animate a hideous creature using a combination of crude mechanical devices and living body parts so that

a human brain that has been kept alive in a glass jar can be given a new body? Yep, this is a prop from one of those movies—the classic brain in a jar. Place this creepy prop on your candy table at Halloween, and those greedy hands will take smaller handfuls after having their appetites challenged by the gooey-looking organ floating in the jar before them. With 20 randomly flashing LEDs protruding from the brain and a handful of wires leading to a "life support" box, you can tell your visitors that your experiments have been a success, and the brain is actually alive and aware of their every move. Add an amplifier or voice changer so that a helper can speak "through" the brain, and you can send your Halloween visitors running back down the sidewalk in fear of becoming unwilling Evil Genius organ donors!

For starters, you will need some large glass jar or fishbowl that can hold an object at least as large as a

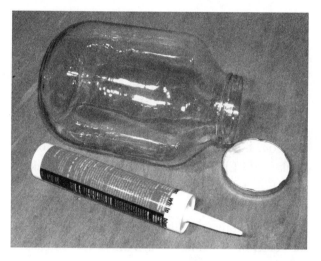

Figure 10-17 *A large glass jar and a tube of caulking*

grapefruit. The opening on your jar will determine the size of brain that you can squeeze into it, so keep that in mind when looking for a vessel. You will also need some type of waterproof material that can be "squeezed" from a tube in order to form what will look like a real brain. I found that white caulking worked great for making the brain, and a single tube as shown in Figure 10-17 along with the large glass jar was plenty to make the brain.

Search on the Internet for some brain photos to get an idea of the object you are going to attempt to model using the caulking, which is perfect for forming the snaking tube-like structures that form the brain. I found it easier to model each hemisphere separately, which would make wiring the LEDs a lot easier, and allow the two halves to be put together inside the jar, resulting in a brain that was twice the size of what could be fit through the opening in the jar I had. Although not the optimal shape, a pair of 4-inch by 2-inch plastic boxes would create a brain about the same size as a human brain when the two halves were covered in caulking and placed back together. It was also easy to mount the 20 LEDs (ten LEDs per hemisphere) to the plastic boxes as shown in Figure 10-18 by drilling two small holes for each LED so the leads could be glued in place. Having a hole for each LED lead kept the leads from

Figure 10-18 *Each box will have ten LEDs*

shorting and made it easy to solder wires to them on the inside of the plastic box.

The LED sequencing circuit will step through 10 LEDs in sequence, but my plan was to wire them randomly so the LEDs would flash with no particular pattern as well as inserting them into the plastic box in a somewhat random position. Each LED sequence will control two LEDs (one on each hemisphere) but, owing to the random order of the LEDs on each side, it will seem as though there were two sequencers running. The LEDs are stuck into the holes in the plastic box so there is enough of each lead to allow a wire to be soldered and then they are glued into position using some hot glue. The LEDs will eventually be held solid by the caulking, so you don't have to use a lot of glue at this point. I also placed the LEDs at random angles and positions on the plastic box, but tried to keep them at the same height so the caulking could be applied easier when forming each hemisphere. The caulking tip is cut so that you can squeeze a line of caulking that is about as round as your index finger with your caulking gun. Try to work around the box so that the caulking forms a continuous but random looking line all around the box like the organic structures in the brain. Try to coat the entire box as evenly as you can, but be careful not to cover up any of the LEDs while you squeeze out the caulking, making what currently looks like a birthday cake with LED candles (Figure 10-19).

When you have made both hemispheres, give the caulking at least a day to set, since you oozed out some pretty large chunks at a time, much more than would be typically used when sealing a bathtub. You can now work on the LED sequencer while the brain dries undisturbed overnight. As mentioned earlier, the sequencer will light up ten LEDs, one after the other, but since the best effect is a random one, the connections between the 4017 decade counter and LEDs is completely randomized. Figure 10-20 shows the LED sequencer circuit, which is a 4017 decade counter driven by a 555 timer that generates the clock pulses that advance the sequence to the next LED.

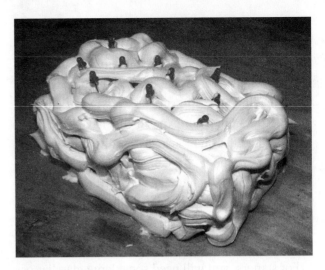

Figure 10-19 *Have you ever had to "make" up your mind?*

Notice how each of the ten decade-counter outputs drive two LEDs, one for each hemisphere.

The 500-ohm resistor that serves as a common ground connection for all the LEDs will also limit the amount of current to the two LEDs that will be on at one time, and you can lower its value if you think your LEDs can take a bit more current, making them brighter. When you are building the circuit on to a bit of perforated board, make room for all the LED connecting wires that will have to reach from the center of the brain, through the jar lid and back to your circuit (life support device). The LEDs will be fine under water, since the resistance of clean water is very low, but the 4017 and 555 timer would not, which is why the circuit is built outside of the brain, which would be very difficult to make watertight. The external box also adds to the "coolness" of the prop, since the 11 wires entering the jar make it look more technical, and you can tell everyone the box is a life support system, or even jazz it up by building it into a larger box with many unnecessary knobs, switches and gadgets to make it look more complex. If you plan to have a hidden helper speak through a microphone to have conversations with your visitors, you might consider installing the speaker into the life support system, to make it seem like the voice is really coming from the brain in the jar. Now to all those wires!

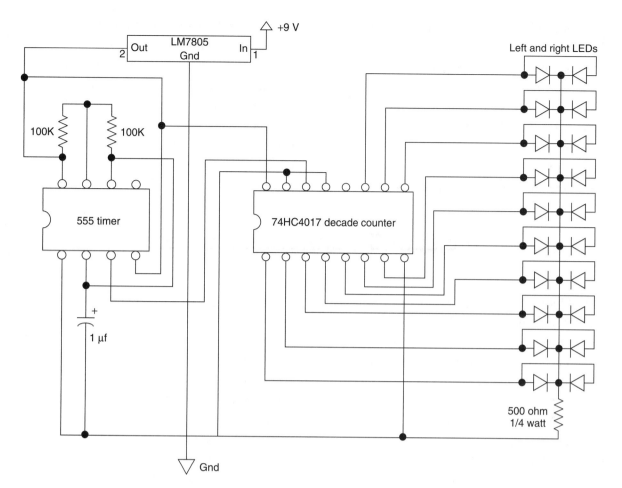

Figure 10-20 *A ten LED sequencer schematic*

After at least one night of curing, the caulking should be rubbery and dry to the touch, so that the brain hemispheres can be handled without a problem. You might also need to take a sharp knife or razor to free the caulking around the base of the box since it has probably set to the surface you put the boxes down on. Solder all of the LEDs as shown in the schematic so that each output from the 4017 drives an LED on each hemisphere, this way you only need to run 11 wires into the jar, 10 for the each LED sequence, and one for the common ground connection. Also, make sure the wires will reach from the bottom of the jar to your life support box as shown in Figure 10-21, or you will have to mount the box on the top of the jar.

The completed brain in a jar (Figure 10-21) actually looked pretty good considering it was

Figure 10-21 *One day this may actually be possible!*

made from nothing more than a pair of plastic boxes covered in bathroom caulking. The LEDs flicker randomly, giving the perception that the brain is busy processing some evil deed, kept alive by nothing more than the space-age device hidden inside the black box. Add a few sound effects or a person with a microphone, and you can have a lot of fun with those who venture close enough to inspect your sinister science experiment. Use your brain (not the one in the jar) to jazz up this project, adding some extra wiring, back lighting, or a handful of useless but complex-looking devices around the jar to make the life support system look more baffling. More switches, more knobs, and more complicated but useless controls. Hey, it works in the consumer electronics market, so it should work for your Halloween displays!

Project 45—Universal Motivator

The next three projects were originally designed as one project, but have been presented as three separate projects because each one can function without the other or in combination with other projects presented in this book. The universal motivator is an easy-to-build mechanical project that will allow you to give motion to your larger Halloween or scare pranks. This device is easily adapted to lift objects, shake trees, open or close doors, move simple animatronic displays and add repetitive movement to just about any prop you can imagine. Because this "motivator" uses a commonly available windshield wiper motor, it can be powered from a battery for installations far from AC power lines and will be capable of pushing or lifting several pounds without any problem. A windshield wiper motor is a small 12-volt DC motor connected to a gear-reducing box so the output shaft will turn between 100 and 300 RMP with considerable torque. If you connected a pair of vice grips to the output shaft of a windshield wiper motor and tried to hold it back while it was connected to a medium-sized 12-volt battery, you would not be able to—they have a lot of power. Acquiring a windshield wiper motor is not difficult or expensive, just go to the scrap yard or auto recycle and remove one from an old car by taking out the two or three bolts that hold it in place. Figure 10-22 shows the dirty old wiper motor I have yanked from some huge gas-guzzling beast at the local scrap yard.

Once you dig a wiper motor out of an old vehicle, it will need to be cleaned up and stripped down to the bare motor by removing all of the wiper arm and mounting hardware. Figure 10-23 shows what the actual motor will look like once you clean up all the grease and remove the support hardware. There will likely be some strange-looking connector on the end of the motor wires, but you can simply cut that off and bare the two or three wires that come directly form the motor. Most wiper motors have two speeds, so there will be a common negative connection and one for each speed. Testing the wire combinations with a 12-volt battery or battery charger is the best way

Figure 10-22 *A typical medium-sized windshield wiper motor*

to figure out which pair of wires will run the motor with the most RPM and torque, and this is done by trying all possible combinations. One pair of wires will make the output shaft run at the highest speed and without any pauses between rotations, so these are the two wires you want to connect. Also shown in Figure 10-23 is the control rod, which will connect to the wipers drive arm.

The wiper motor will need to be mounted to a sturdy platform made of wood or metal with enough room to carry whatever 12-volt battery you plan to use. The battery will also become the weight that will hold the board in place, and should be no smaller than a motorcycle or recreational battery and have a rating of 12 volts. Windshield wiper motors are very powerful, and require a minimum of a few amps just to start turning, so a 9-volt battery is not going to cut it here. If you ended up with some of the mounting hardware still attached to the wiper motor like I did, then see if you can reduce it to the minimal size and shape needed to securely fasten your wiper motor to the board as I did in Figure 10-24. If you look back to Figure 10-22, you can see how the original mounting hardware was cut down to what is essentially an L-shaped bracket to bolt the motor to the board. If you don't have any mounting hardware, you will have to make your own motor bracket from some shelf-mounting hardware or a bit of angle iron by drilling holes for the motor shaft and mounting bolt holes.

With the wiper motor cleaned, tested and mounted to a sturdy platform, all you need to do is add your battery and a way to switch on the power and you will now have a mechanical motivator to move your Halloween displays. A pair of quick-connect terminals in place of an on–off switch will allow you to run a long wire from the motivator to your "control center," so you can run multiple displays from a single location well out of view. The simple schematic shown in Figure 10-25 will allow any type of pushbutton, flip switch or automatic activator (shown next) to engage the motor, adding life to whatever it may be connected to. Polarity of the battery is not important here

Figure 10-23 *The bare wiper motor cleaned and tested*

Figure 10-24 *The wiper motor mounted on the wooden base*

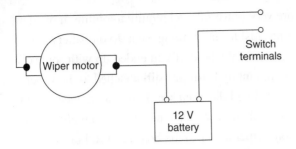

Figure 10-25 *Wiring diagram for a remotely located switch*

Figure 10-26 *The completed universal motivator*

Figure 10-27 *A steel rod adapted to the motivator*

because the motor movement is reciprocating (back and forth), and will not be any different no matter which way the motor shaft is rotating.

The wire you will use should be at least as heavy as a typical AC electrical cord, since the wiper motor will draw between 5 and 20 amps depending on how much you load it down. If you try to use the thin wire you normally use for low-power projects, then you may be adding a bit of smoke to your Halloween show as the wires get hot, like toaster coils from a current overload. Figure 10-26 shows my completed motivator, including a 30-amp hour wheelchair style lead acid battery and the heavy-gage wire used to connect the motor to the power source. To start the motivator running, the two quick connect terminals simply need to be shorted together, which is done manually via a remotely wired on–off switch, or by some automatic activator like the one presented next.

The actual movement generated by this motivator will be transferred to your display or prop by either a rod or a wire connected to the output arm of the gearbox connected to the wiper motor. This arm is typically 3 inches in length, so you will end up with a total movement of 6 inches as the arm makes one complete revolution. Six inches of movement is a lot, especially considering the amount of push or pull this unit can deliver, so your Halloween displays can be made large and heavy for maximum viewing impact. To connect a rod to the motivator, cut one end from the original wiper control arm and bolt it to whatever length of rod you need. This rod can be made from wood,

plastic, or metal like the one shown in Figure 10-27. The rod can be interchanged by simply removing the two bolts and replacing it with another rod or some other device that will deliver the movement from the wiper's arm to your prop or display.

The bolts used to connect the rod to the wiper arm shaft should not interfere with the mounting hardware or your electronics when the motor makes a complete revolution, so test this before setting up your display. The end of the original control arm that is now connected to your steel rod will be held to the wiper's drive arm by some type of easy-to-remove C-clamp or a cotter pin, which will prevent it from falling off as the motor rotates. Figure 10-28 shows the rod and all supporting hardware connected and ready to push and pull whatever may be connected to the other end of the steel rod. Some extra weight besides what is offered by the heavy battery may need to be added to the platform as well, especially if your target prop is

Figure 10-28 *Motivator ready to push or pull a large object*

large and heavy. The windshield wiper motor is more than capable of dragging a 50-pound weight around, so you might want to consider an even heavier

mounting platform made of steel or with enough room for several bricks for ballast.

With the universal motivator, you can add life to your Halloween displays, making huge bats that flap their wings up and down, ghosts that wave back and forth, witches that constantly stir the cauldron, or even indoor displays like my flesh-eating Jack-O-Lantern presented in Project 47. The motivator can be further expanded for fully automatic operation by inserting a relay-operated circuit in place of a manually operated switch on the motivator's output jacks. Presented next is a very simple device that will switch on the motivator any time a sound is played, so you can sync your prop to some pre-recorded music or sound effects, making it jump to life without having to flip a switch on or off manually.

Project 46—Sound-activated Switch

If you have your entire yard rigged up on Halloween like a horror movie set, then you won't need a way to activate all your props and gags automatically; you can just sit back and enjoy the look on your visitors' faces as they flee from your yard in a panic! The universal motivator presented in the last project is a perfect candidate for automatic sound control, so you can simply connect it to your prop and let your music or sound effects trigger it when needed. The sound-activated switch presented in this project can also turn on lights and many other AC or DC loads any time there is sound fed into it, and it can even do a bit of synchronization depending on the nature of the sound. The sound-activated switch schematic presented in Figure 10-29 is simply a relay driver that energizes the relay's electromagnetic coil when the sound level fed into the base of the transistor reaches a pre-set level. If you feed the output from an MP3 player or computer sound card into the

device, whatever load is connected to the relay will be switched on when the sound begins or reaches a certain level. By adjusting the level of the 100K variable resistor, you can even make the relay respond to drum beats or heavy bass in a music or sound file.

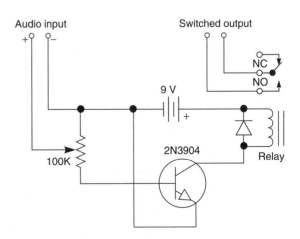

Figure 10-29 *A sound-activated relay driver*

Figure 10-30 *The sound-activated circuit completed*

The sound-activated switch runs from a 9-volt battery for many hours, and can switch on whatever load the relay you use has been rated for. A typical 5-amp relay will be large enough to activate the universal motivator presented in the last project, so you can have your displays automatically respond to your sound effects. Like any relay driver, the generic diode is placed across the relay's coil to stop back currents from damaging the NPN transistor. The parts are placed on a small perforated board as shown in Figure 10-30, with a power switch and external variable resistor for easy setup. I also added a ⅛ mono jack for the sound input so I could use commonly available cables to connect the sound-activated switch to my MP3 player or computer's audio output.

The circuit board and battery are placed in a plastic box as shown in Figure 10-31, completing the sound-activated switch. The two terminals on the front of the box are the same type used to switch on the wiper motor in the last project, and they will often be connected directly together so my music can operate whatever is connected to the motivator. To set the sound level on the audio source and sound-activated switch, turn up the 100K variable resistor to the maximum position then slowly turn up the volume on your source audio device until the relay begins to click. It does not take much output to drive the relay, so your MP3 player or audio device may only need to be at 2 or 3 percent volume level. If you find that the relay is always on, then that's

Figure 10-31 *The sound-activated switch*

when you will have to lower the setting of the 100K variable resistor to reduce the input sensitivity. The variable resistor is also handy if you want your audio source to share the sound-activated switch with an amplifier through a splitter cable so your visitors can hear the sound as well.

If you find that this simple sound-activated switch works well with your mechanical displays, then you may even consider building a multi-channel activator so you can sync your entire Halloween show with one or more audio sources. Any stereo sound player will offer two independent audio channels, so you could edit a "control file" on your computer using sound-editing software, which would allow precise triggering of two synchronized channels. How about running a microphone into a voice changer connected to the sound-activated switch to make the jaws of a flesh-eating Jack-O-Lantern chomp up and down as you recite evil phrases into the microphone? Yes, the possibilities are endless with the last two projects combined, and I really like that flesh-eating Jack-O-Lantern idea, so I am going to show you how I pulled it off in the next project!

Project 47—Flesh-eating Jack-O-Lantern

A lot of people have a Jack-O-Lantern at their doorstep on Halloween, but only a true Evil Genius has one that will try to attack your guests. Yes, indeed, by using the universal motivator and the sound-activated switch presented in the last two projects, you can bring that boring old vegetable to life with a vengeance. Normally, I would use a real pumpkin for something like this, inserting a hinged metal brace between the two halves to make a jaw, but dude, we wrote this book in the spring and our pumpkin plant was only an inch high, so we had to make one from some parts found at a hardware store. You could also use one of those plastic Jack-O-Lanterns, but they always seem to look too happy for my taste. Evil—I need evil!

Let's start by remembering some of the art we made in public school using a messy yet inexpensive technique called paper maché, a simple technique that is actually great for making masks,

piñatas, props, and various hideous dead things. With some paint, flour, and newspaper strips, you can actually sculpt some very decent-looking large props around wire or balloons ranging in size from fist size to life size. I wanted a large Jack-O-Lantern with shark-like teeth that looked like it could swallow half an adult in one bite, so I went on a tour of a local hardware store looking for an inexpensive round thing that could be used as the body. I found various pots and planters in the gardening department that were about the right size, including the two 24-inch diameter hanging planters shown in Figure 10-32, which had solid wire frames and some kind of thick liner material that would be perfect for covering with paper maché.

Because I wanted the two planters to form a hinged jaw, allowing the Jack-O-Lantern to chomp its evil teeth, I had to install a hinge between the two halves by welding it to the metal screen as

Figure 10-32 *These two planters will form a sphere*

shown in Figure 10-33. A welder comes in handy for these small jobs, but if you do not have one, then fastening the hinge will have to be done with bolts and large washers or tubing clamps. If you plan to use similar planter frames, be mindful of the position of the wire as you orient the hinge so that the area you plan to cut out for eyes later will not be obstructed by the wire frame. The lightweight wire frames should open and close with ease once the hinge is secured to each half.

The mesh-like liners that came with the planters would also come in handy, creating most of the body material, so that only a few layers of paper

Figure 10-33 *Hinging the two wire-frame halves*

maché would be needed to make a rigid covering. I had to trim the liners down and tie wrap them to the frames so that the two halves would close, but other than that, they were perfect. It's too bad the liners would not fit around the outside of the globes to form the outer skin, but they were just too small to stretch around, so they will have to become interior liners. Figure 10-34 shows the trimmed and fastened interior liners held to the screen globes using a few tie wraps.

Now for the messy fun, adding a paper-maché skin to the outside of Jack's body to hide the wire screen and create a paintable surface. This process can be done with regular flour, wallpaper paste, and many other combinations of inexpensive materials, so hit Google up for "paper maché" to find out which method you might prefer. As my evil helper Kathy demonstrates (Figure 10-35), you basically dip newspaper shreds into a mixture of gooey baking flour mixed with water, run off the excessive mixture between your fingers, place the gooey paper strips on to the object you want to skin, then let it harden overnight into a rigid and paintable surface that can be cut or added to later. For a more rigid final product, use three or more layers of newspaper, and for single-use objects that will not be handled, two layers will usually suffice. The wet newspaper strips are worked into each other with your hands

Figure 10-34 *Interior liners trimmed and fastened*

Figure 10-35 *The messy art of paper maché in progress*

Figure 10-36 *Anyone want to read the globe?*

until the entire surface is one big sloppy mess, covered in paper strips and paper maché goop.

Figure 10-36 shows the final result of a two-layer paper maché covering after a night of drying. The final product is actually pretty rigid considering it is made out of nothing more than newspaper strips and some baking goods. I took a sharp knife to trim around the globe halves to allow the two parts to be opened once again, then added a ring of duct tape around the "lips" just to strengthen that area before painting. Eyes were also cut out of the front using a sharp knife and then a ping-pong ball cut in half

was taped behind each eye socket to make Jack look a little creepier. If your paper is still a bit damp, then let it dry out in the sun or under a hot lamp until it is completely dry and rigid to the touch, and if the final skin does not seem thick enough, you can always add a few more layers any time to make it thicker.

I painted Jack's head with some pumpkin-orange spray paint and then bolted him to a large plywood base by screwing a small piece of wood down through the inside of the lower globe to hold it there like a vice. The same type of "wood sandwich" method was used to make a mounting place for the universal motivator end point at the back of Jack's head as shown in Figure 10-37. An L-shaped bracket was then screwed to the wood plate at the back of Jack's head so a bolt could connect the business end of the motivator's push and pull rod to the top of Jack's head, allowing his mouth to open and close under motor power. The rod mounting block was placed high enough on the back of the top globe to get sufficient leverage and allow the mouth to open and close approximately 6 inches with each motor revolution. The position of the universal motivator will also determine how far Jack's mouth will open and close, since this will limit the movement of the push and pull rod, making it easy to keep the mouth open a little bit at all times.

Figure 10-37 *The universal motivator rigged to Jack's head*

Figure 10-38 *Feed me! Chomp, chomp, chomp!*

Once Jack was flapping his flesh-eating teeth in an acceptable manner, I added the sound-activated switch to the universal motivator so I could play a pre-recorded message by pressing a single button, which would also control the movement of his mouth. It was also fun to speak into a pitch changer fed into the sound-activated switch to make the experience a little more interactive to trick or treaters that came to the door. A nice long evil laugh would make Jack's huge shark-toothed mouth open and close in sync with the sound, making it seem as though he were coming to life, ready to swallow anyone brave enough to venture up to the door. As shown in Figure 10-38, Jack the flesh-eating Jack-O-Lantern was an evil-looking beastie with enough room in his mouth for an entire bag of candy, and maybe an adult head as well! Hey, it's all about evil on Halloween, right?

Well, there you have it, more evil deeds done dirt cheap. When you put your mind to it, just about any old junk and the most unlikely materials can come together in the Evil Genius lab to create works of art, scary hideous art perfect for scaring away those who would come to your door on Halloween to take away all your treats. Don't let this wonderful night fall victim to smiling skeletons and happy ghosts, put the evil back into Halloween with your Evil Genius imagination!

In the next section, you will learn how to make the ultimate high-tech prank and meet our furry critter, Fluffy. Get the video camera ready to capture hilarious shrieking, running, and jumping!

Chapter 11

Fluffy Attacks! Scare Them Silly!

This project is by far one of the most effective and hilarious Evil Genius pranks I have ever made, and although it is a bit more involved than the other projects presented in this book, I assure you, the results are definitely worth the effort. Fluffy is a furry critter about the size of a skunk who has been trapped in a wooden cage, and like all caged animals, wants to spring free and unleash a surprise attack on his unsuspecting captors, which he indeed does with explosive action. Fluffy will sit motionless in his cage as his captors silently approach to inspect their catch, but as they get closer, he will burst from his cage, throwing the trap doors wide open as he flies through the air at a height of 5 feet or more in a surprise aerial attack.

As Fluffy is triggered by a bright light, his trapdoor cage can be placed outdoors, in a garage, or any other dark area where a critter might be invading, making it seem as though he has finally been subdued. As you slowly creep up to Fluffy's cage with a flashlight in hand, whisper to your unsuspecting buddy that you think your trap has finally caught the critter that has been terrorizing the neighborhood. Shine the light through the wire mesh in the realistic looking trap to reveal the black furry body that lies motionless in the back of the cage as you try to creep up for a closer look.

Just as your buddy is in the "target zone," hit the front of the cage with the beam of light to trigger the light-sensitive launch mechanism and send Fluffy springing from his cage in an explosive burst of energy, directly at your unsuspecting pal who will soon be cowering in fear or letting out the magical "little-girl screech." Fluffy's launch system can also be manually triggered for close-range "show and tell" displays where several people might be crowded around his cage, unknowingly becoming targets of a flying furry attack.

Owing to the realism involved in the cage and the explosive manner of Fluffy's launch, this prank is guaranteed to scare the wits out of anyone caught in his attack path, leaving the most stone-faced, scare-proof "tough guy" running for cover on the nearest table or chair! Because this project is more involved than many of the pranks in this book, it is broken down into several parts so you can make changes to suit your own evil agenda or simply work with parts you already have on hand. There is plenty of room for modification or improvement on my original design, but it is probably a good idea to read through the entire project to see how all the parts work together before deciding on any radical modifications.

Fluffy's spring-loaded launch pad is by far the most important part of this project, and all the other components must be built around it, which is why it is built first. To be effective, the launch pad must be capable of hurling an object the size of a soup can to a height of about 5 or 6 feet, with enough force to blow open a pair of small wooden trap doors. For this purpose I acquired a pair of pull springs from a hardware store with a length of approximately 6 inches, and a diameter of about a ½-inch. Although these springs can be pulled apart by hand, they do offer a lot of resistance, and can be easily mounted to a bolt due to the rounded ends. As shown in Figure 11-1, you will also need a pair of door or gate hinges and some ½-inch-thick plywood to make some of the launch pad parts.

If you plan to follow my design closely, then the launch pad and cage base is made from a piece of ½-inch-thick plywood cut to a length of 18 inches and a width of 12 inches. The actual launch pad is cut to a size of 7 by 11 inches and is also made of the same ½-inch-thick plywood.

The launch pad is basically a movable paddle held to the base by the two door hinges and pulled upwards by the spring to create the launch force. If you plan to make a much larger critter and cage, then your launch pad will have to be at least as wide as the critter you plan to hurl and approximately two or three times in length. For a much larger launch system, use thicker plywood as

well, or you will find the wood bending under the tension of the springs. As shown in Figure 11-2, the launch pad is held to the launch base by the two door hinges so that there is approximately 1 inch of clearance between the end of the launch base and the top (unhinged part) of the launch pad. The launch pad should easily travel from the down position to more than 90 degrees with very little resistance. Also note that it may not be possible to flatten the launch pad completely against the base due to the hinge screws on the underside of the launch pad, but this will not be a problem.

The block of wood shown on the other end of the launch base in Figures 11-2 and 11-3 is just a bit of 2 × 4 cut to a height of 3 inches to support the launch springs. This spring support block is held to the end of the launch base by a pair of wood screws and will easily take the force of two or more springs without a problem. The wood screws used to hold the 2 × 4 to the wood base should be at least 1.5-inches long for strength, and you may want to pre-drill the hole with a drill bit approximately half the diameter of the wood screws to prevent the 2 × 4 from cracking as you tighten the wood screws. The two springs are then fastened to the top of the spring support block as shown in Figure 11-3, using the same 1.5-inch long wood screws and a few washers or a bolt to keep the spring's body away from the surface of

Figure 11-1 *Two pull springs and a pair of hinges*

Figure 11-2 *Fastening the hinged launch pad to the launch base*

Figure 11-3 *The launch springs are mounted to the support block*

Figure 11-4 *Launch springs mounted to the launch pad*

the 2 × 4 block. Without the washers or bolts for spacers, the springs would be compressed against the wooden block, and might fatigue or break over time. This system allows the springs to move up or down, pivoting on the screws that hold them to the block as the launch door is loaded and released.

The proper position to mount the launch springs to the launch pad will depend on the length and strength of the springs you are using. In my system, the springs worked perfectly if the mounting bolts are placed 2.5 inches up from the hinged end of the launch pad as shown in Figure 11-4. When at rest, the door sits at an angle of approximately 45 degrees, and is under some serious tension when forced down against the wooden base. Your launch

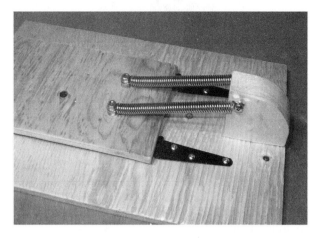

Figure 11-5 *Launch springs in the loaded position*

pad should be pulled to at least 45 degrees by the springs when at rest and allow the springs to stretch apart when placed flat in order to launch the critter effectively. To find the optimal position to bolt the springs to the launch pad, just lower the launch pad all the way, then pull one of the springs by hand along the pad until the coils start to separate; this is the correct position of the spring when the launch pad is laying down.

The springs are held at the appropriate position on the launch pad using a pair of ³⁄₁₆-inch nuts and bolts with the appropriate washers for spacers just as it was done on the 2 × 4 end of the springs. The washers will prevent the body of the springs from being crushed against the launch pad, which would cause the loops at the ends of the spring to break eventually. With the two launch springs fastened to the launch pad as shown in Figure 11-5, you can now test the launch force of your system. Place a hockey puck on the launch pad or a baseball, then pull back on the launch pad until it is as close to the wooden base as it will go, releasing the payload to see how far it will travel. A small lightweight object should easily fly across the room at a height of 6 feet or more, but keep in mind that this distance will be slightly reduced once the overhead trap doors are also placed into the equation.

The furry critter in my design is just a large soup can covered in fur, and it will easily make a 20-foot journey at a height of 10 feet or better when hurled directly from the bare-bones launch pad shown in Figure 11-5, which would be overkill for this prank. Because of the trap-door design of the completed cage, this distance is effectively cut in half, so keep that in mind as you tweak the springs for optimal performance. Oh, and be careful when playing around with the loaded launch pad—rumor has it that it really hurts when you swat yourself in the chin by accident! Once your launch pad is working the way you like it, you can move ahead and build the rest of Fluffy's cage.

Project 49—Trap Door Cage

With the launch pad working correctly, the rest of the cage is simply a prop built around the base of the launch pad to add realism to the prank. The top of the cage will also contain the barn door-style trap doors that will be forced open as Fluffy begins its high-speed journey into the air like a World War I flying ace on a kamikaze mission. You can let your artistic woodworking skills run wild on this part of the project, but do keep in mind that the position of the trap doors above the launch pad must be accounted for in order to achieve an effective successful launch. From this point on, almost all of my cage will be made by cutting up some 1 × 2 to create what looks to be an old-fashioned animal trap built by some scary old codger with a knarly gray beard who lives in the back woods and dines on whatever small critters are unfortunate enough to cross his land (yikes)! Anyhow, the sides of the cage are made by cutting eight lengths of the 1 × 2 to a height of 10 inches, and then the top and bottom frames will be made from four pieces of the 1 × 2, which will form a rectangle exactly the same length and width as the launch-pad base (18 × 12 inches). Figure 11-6

shows the 1 × 2s cut to the appropriate lengths as described. I chose the rough-cut 1 × 2 for its rugged look, but feel free to make the rest of the cage from whatever scraps of wood you may have on hand.

The length of 1 × 2 used to make the cage frame sides are simply nailed to the corners of the cage after the bottom frame has been installed under the launch-pad base as shown in Figure 11-7. The frame under the launch-pad base creates a very sturdy foundation and offers plenty of room to hide the electronics, launch mechanism, and batteries under the launch-pad plywood base. Don't worry about doing a professional woodworking job on this cage, since it will look more authentic if it is a bit rough and unfinished, just like a makeshift animal trap would be.

The top frame of the cage is made exactly like the base, so that the outer dimensions of the frame are the same as the dimensions of the launch-pad base. The top frame is then installed flush with the top of each of the side supports as shown in Figure 11-8 to create the rest of the cage frame. To keep your vicious rabid animal from escaping, cut and staple some wire mesh along the open sides of the cage, again adding to the illusion that the trap may indeed be the real deal. Now the cage lid will be made, which will serve as the top of the cage as well as becoming a bumper for the traveling launch pad.

The top of the cage is just another piece of ½-inch plywood, but its dimensions will determine the launch angle of the critter as it leaves the launch pad because it will become a stopper for the launch pad. When at rest, my launch pad is held at an approximate 45-degree angle, but can

Figure 11-6 *1 × 2s used to form the rest of the cage frame*

Figure 11-7   *Cage base and side supports installed*

Figure 11-8   *Cage frame and screening completed*

travel much further when reaching its full range of motion under tension of the springs. If Fluffy were to be lunched without some method of restricting the launch-pad travel, the little critter would most likely be slammed towards the ground rather than into the air as desired. To send Fluffy into the air at an appropriate eye-level height at a distance of about 10 feet from the cage, the cage lid is cut so that the launch pad cannot travel more than about 80 degrees. You may want to experiment with the

best launch-pad breaking position by temporarily nailing the cage top in place at different positions while you try a few test launches in order to get a good height and distance. As shown in Figure 11-9, my cage lid will stop the launch pad at about 80 degrees, and is cut to a size of 12 × 6 inches.

When the launch pad slams into the cage lid, it also makes a loud bang, but this just enhances the overall shock to the innocent bystander who will soon be jumping up on to the nearest piece of

Figure 11-9 *The cage lid is also a launch-pad brake*

Figure 11-10 *Trap doors installed on hinges*

furniture to avoid the ankle biting that they fear will be coming next. Now all that is left to creating Fluffy's cage is the two trap doors that will reveal his furry body and pop open when the launch pad sends him on his way. The two doors are cut from the half in plywood so they are just large enough to cover whatever space is left open on the top of your cage. They are then held to the top frame by a pair of small hinges so there is little resistance

when they need to be flung open. Each of my cage trap doors is 6 × 9 inches in size and held to the top frame with a 2.5-inch door hinge as shown in Figure 11-10. The corners of the launch pad should strike the trap doors somewhere near their midpoints in order to achieve an effective launch.

To test the launch system now, you will have to poke a screwdriver through the screen at the rear of the cage as you press down on the launch pad to

Figure 11-11 *Completed and painted cage*

set it in the loaded position. Use the screwdriver blade to hold the launch pad down as you place a glove or ball on the launch area then close the two trap doors. Quickly yank the screwdriver off of the launch pad to send the object into orbit and test the efficiency of your newly installed trap doors. The object should travel at least half as far as it did with the doors open, but now the effect will seem much more explosive as the two trap doors are flung open at lightning speed. If your trap doors were also sent to the other side of the room, then dude, either your hinges were too small, or your springs were too big! Again, a little

experimentation might be needed in order to get things working the way you want, but soon you will be watching your buddies run for cover. If everything seems to be in order, it might be a good time to paint your cage if you think it needs it. As shown in Figure 11-11, I gave my cage a dark brown paint job using a spray can, which makes it more difficult to see what is hiding inside while looking through the wire mesh.

Now you can move on to the next step in Fluffy's evolution—the light-activated or manually controlled launch mechanism, which will help him become airborne.

Project 50—Light-activated Trigger

Fluffy's triggering mechanism must hold down the spring-loaded launch pad until the innocent bystander (victim) is standing in the correct place for a flying attack from the furry beast. The trigger can take the form of a simple hidden switch placed at the rear of the cage that the Evil Genius (you) can operate while the victim is busy peering into

the cage at a safe distance or it can be a fully automatic light-activated trigger that responds only to a direct beam from a flashlight so you can work it from a distance. The light-activated system is much more fun because you can become part of the prank, creeping up on Fluffy as if you are actually afraid of what might be caught in the cage as you lead your unsuspecting buddy along with you. Because you are also in the line of fire, there will be less suspicion than if you were simply trying to sneak your hand behind the cage to press the trigger button. No matter which method you decide to use, including one of your own design, you will need the same mechanically activated solenoid actually to release the loaded launch pad, so you must find a way to hold the loaded launch pad in place so that it can be easily released with approximately ⅛ inch of mechanical motion. As shown in Figure 11-12, a simple L-shaped shelf bracket is fastened to the underside of the launch pad using a couple of wood screws so that, when the launch pad is pushed all the way down, one end of the shelf bracket can drop through a hole in the base of the cage into the underside where the solenoid and electronics will be hidden.

The shelf bracket I used is 1.5-inches long on each side so it can easily fit through the hold in the cage base to allow a pin to lock it in place through one of the holes in the bracket while it is on the underside where the solenoid will be located. The shelf bracket should enter the hole in the cage base without any resistance, so drill a hole a bit larger than the width of the metal bracket. The easiest way to get the hole in the correct place in the cage base is to first install the shelf bracket to the launch pad and then press it down to mark the correct hole position. If you do not yet understand

how the shelf bracket will become a locking mechanism for the launching pad, then read ahead a bit to see how everything gets put together. With the shelf bracket installed so that it can lock the launch pad to the underside of the cage base, you will now need to dig into your electronics scrap pile for a suitable electromagnetic solenoid. As shown in Figure 11-13, I found a suitable electromagnetic solenoid and bolted it to a bit of angle iron so it can be mounted to the underside of the cage base.

A suitable electromagnetic solenoid will have a plunger that pulls inward when power is applied to it from a 12-volt battery back. The solenoid does not have to be rated for 12 volts, but it should have a decent amount of pull when running from a 12-volt power source in order to reliably pull the release pin. My solenoid is actually rated for 24 volts, but it pulls in with a fair amount of force on only 12 volts, enough so that it is very difficult to pull out the plunger once it retracts all the way into the coil. The plunger will need to travel only one-eighth of an inch from its fully retracted position, which is why a solenoid rated over 12 volts will probably work just fine with less power. I have even tested a solenoid rated at 120 volts AC and it worked fine on a single 12-volt DC power source. To test your solenoid's pulling force, pull the plunger back about ⅛-inch from the fully retracted position then try to hold it back while you apply the 12–volt power source. If you cannot hold it

Figure 11-12 *A launch-pad locking system made from a shelf bracket*

Figure 11-13 *An electromagnetic solenoid and plunger*

back without a lot of force, it will probably have enough pulling power to release the triggering pin reliably every time. Now you will need to make the pin that will hold the loaded pad in place and rest on the end of the solenoid plunger so that the plunger only needs to retract approximately one-eighth of an inch in order to make the pin fall out of the hole. This process may sound complicated, but it is extremely easy to implement, and works much like the way a typical mousetrap works. As shown in Figure 11-14, this locking pin is just a bit of metal that can fit through the shelf-bracket hole while it rests on the solenoid plunger when it is pulled only slightly from its housing. I made my locking pin by filing down a piece of another shelf bracket, but you could make it from any bit of scrap metal you have laying around or even from an old kitchen utensil (fork prong).

Also shown in Figure 11-14 near the hole where the shelf bracket enters the cage base is another bit of a shelf bracket fastened to the wood so the locking pin rests on it instead of directly against the wood. This bit of metal is optional, but it does stop the locking pin from eventually wearing a hole in the wood as it is released in a violent manner when the solenoid pops its plunger out of the way. To get the solenoid in the correct position so the locking pin will fit optimally, load the launch door,

Figure 11-14 *The locking pin holds the loaded launch pad*

and place the pin through the hole, then hold the solenoid in place and mark the screw holes. Now you can carefully load your launch pad by forcing it all the way down and placing the pin through the hole on the shelf bracket attached to the launch pad while you let the other end of the locking pin rest under the ever-so-slightly retracted solenoid plunger. Keep in mind that the loaded cage is like a mousetrap—too much movement or jarring could set it off and, if your hand is in the way, you will remember not to do that again. You now have the option of simply wiring a switch from the solenoid directly to the battery pack for manual operation, or adding a light-triggered switch to allow covert remote triggering of the solenoid. Regardless of the method you plan to use, it's a good idea to start with the direct switch method first, since it will allow easy testing of the entire system, and in the end, you will need a master switch anyhow.

For proper solenoid operation, you will not get away with a 9-volt battery, or even a pair of 9-volt batteries wires for 18 volts, as there is simply not enough current available in such a small battery, resulting in a very weak solenoid pull. A large 12-volt flashlight battery or a 12-volt pack made from eight 1.5-volt flashlight batteries will be needed to get this project up and running, and should yield hundreds of launches without a problem. I chose the multiple D cell battery pack approach as shown in Figure 11-15, and had plenty of room to spare once the battery holders were screwed to the underside of the cage base. The direct switch method is so simple that I will not bother with a schematic diagram, just wire the switch in series with the battery pack and solenoid so it is activated when the switch or momentary pushbutton is flipped to the on position. With a good 12-volt alkaline power source, the solenoid plunger should have no problem at all retracting back into the coil to release the locking pin. Have you spanked your hand with the flying launch pad yet? No? You will in due time.

The light-activated system is a bit more complicated than the single switch method, but not

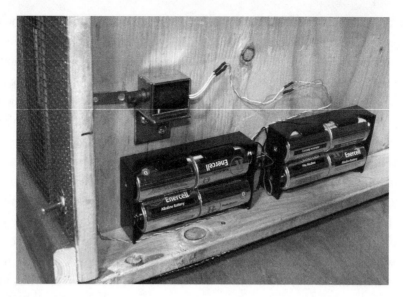

Figure 11-15   *Installing the battery packs under the cage*

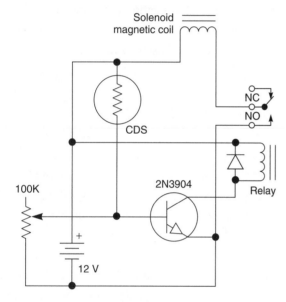

Figure 11-16   *Light activated trigger schematic*

by all that much. As shown in the schematic diagram (Figure 11-16), a common 2N3904 NPN transistor has its base connected to a light-sensitive photocell so that any bright light source will saturate the base of the transistor, switching on a small relay, which will in turn send all of the available power to the solenoid coil. The 100K variable resistor is used to set the sensitivity of the photocell so it can be used in a light or dark room without a false triggering due to ambient light sources. The diode across the relay can be any commonly found diode and is simply there to

protect the transistor from any voltage spike induced by the relays coil. The reason the transistor is not connected directly to the trigger solenoid is because there will not be enough current switching capacity in the small transistor to pull in the plunger with adequate force. By using the relay to switch the power to the solenoid, all of the battery pack's current will be available to the solenoid coil. Also, if you are not yet familiar with the CD cell (photocell), then have a look back at Chapter 3, Project 6 for a good description of how they work and where you can find one in a hurry.

The light-activated circuit can be built in a small bit of perforated breadboard with the appropriate length wires for the photocell, sensitivity adjustment potentiometer and power supply as shown in Figure 11-17. You could actually mount a small variable resistor directly to the circuit board, but for fine tuning of the light sensitivity in a bright room, it's much easier to have access to the variable resistor without having to tip over the cage. In most cases, any setting of the variable resistor will be fine when using a flashlight to trigger the launch circuit in a dimly lit room, but in a room with a lot of ambient light, you may need to tweak it a bit to avoid false triggering. The power switch used for testing the directly connected power source will still be used to switch

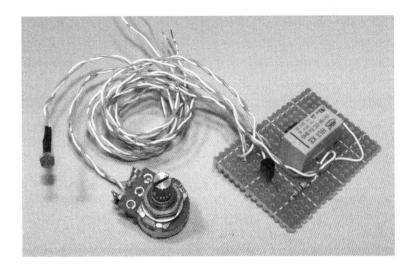

Figure 11-17   *Completed light-activated trigger circuit*

Figure 11-18   *Installing the circuit board and photocell*

the battery pack to this circuit because, over time, the transistor will eventually drain the batteries due to the tiny amount of current leaking through the emitter, even when the room is dark. The best way to set up this prank is to adjust the sensitivity ahead of time, then switch on the unit just before your victims have arrived for their fright session. You can leave the unit running for an hour or two without any real drain on the batteries, but I wouldn't leave it on for days at a time unless you really can't help it.

With the circuit board completed, bolt it somewhere under the cage base and then run the wires to the appropriate places so the master

switch and sensitivity adjustment are on the rear of the cage and the photocell is on the front of the cage. As shown in Figure 11-18, the photocell is placed halfway into a hole drilled on the front of the cage, so it is almost unnoticeable, yet very sensitive to any direct bright light such as the beam from a flashlight. Placing the photocell only halfway into the hole also blocks a lot of ambient light from any bright room light that could cause a false triggering.

With the circuit board and photocell properly installed and fully wired, test the unit by turning on the master power switch and using a flashlight to trigger the system. Rather than loading the launch pad every time, just listen for the relay to click as you mess around with the sensitivity adjustment and positioning of the cage in your target area. Once you have everything set for the room you plan to use, then carefully load the system with the master power in the off position. Did you smack yourself with the launch pad yet? Trust me, it's coming! Now you are almost ready to test your buddies' fear of small flying rabid animals. All that is left is the actual critter's body and a way to hold him in place before launch.

Project 51—Fluffy's Body

You can launch just about anything from this cage (I can hear your evil gears turning), but the most obvious thing would be a small hairy animal—after all, it does look like an animal trap. Extreme detail is certainly not important here because the attack happens with such a loud bang and at such a high speed that there is only time to duck or scream, not enough time to identify the species of critter that is heading for your neck at high speed. Also, you don't want to launch a heavy object or it will become a weapon rather than a prank, and most likely not travel very far before thudding to the ground. So far, the best critter I have sent into orbit has been the most basic design—a large soup can with some fuzzy black hair glued around it. This critter also behaves very well when sitting on the launch pad because it is held in the perfect position using a magnet fastened to the launch pad. The magnet shown bolted to the launch pad in Figure 11-19 was taken from a dead computer hard drive and is very strong considering it size and weight. This magnet easily holds the metal can in place, even when covered in fur, but due to the explosive force of the launch system, does not reduce the overall range of Fluffy's flight. You can use any small magnet that will effectively hold a soup can in place, but do make sure it is well fastened to the wood, or you will be launching more than a can of fur.

Also, do a few test launches before you permanently fasten the magnet to the launch pad so you can make sure the critter does not strike the cage lid or sides on the way out of the cage.

As shown in Figure 11-20, I removed the contents from the can by cutting away only a small section of the can lid with a pair of pliers so it would remain rigid and allow an area to glue the fur. At this point, I can lock and load the launch system, and trigger the can to fly halfway across a large room by hitting the photosensor directly with the beam of a flashlight. The cage is aimed so that the can flies just to the side of where I plan to have my victim stand so that there are no "head-on" collisions between human and beast.

Now it's time to give Fluffy his fur. An old furry mitten or scarf could be butchered for the fur, or you could head to the local fabric store and ask the nice salesperson for a small chunk of fur that looks like a rabid mangy rat that has been dragged through a cesspool of toxic waste (nice). I chose the new-fur approach and simply cut a piece of

Figure 11-19 *A hard disk drive magnet holds the can in place*

Figure 11-20 *A great way to deliver canned goods to your friends*

Figure 11-21 *Extreme soup can makeover begins*

fabric large enough to wrap around the entire can as shown in Figure 11-21. Remember, accuracy and detail are not important as long as the thing that flies out of that cage does not look like something you would want rustling around in your shorts, so keep it simple.

A great way to keep the fur securely fastened to the can so it will survive the blast through the trap doors as it leaves the cage is by using a hot-glue gun to lay strips of glue about 2 inches apart around the entire can. As shown in Figure 11-22, a strip of glue is added to the can and then the fur is

rolled over the glue to fasten it permanently to the can. Also, make sure to cut a piece of fabric large enough to wrap the entire can and fold in over the sides to hide any of the shiny reflective metal can. Once you are done wrapping the fur around the can, add the hot glue to the ends of the can and bunch it up around each end to completely bury the can. Fluffy is now ready for attack.

If you really want to go overboard on the detailing, you might consider gluing a pair of those cheesy plastic fangs and some googlie eyes to Fluffy's body just to be silly, although I doubt

Figure 11-22 *Hot gluing the fur to the can*

Figure 11-23 *12 ounces of pure fear-invoking critter*

Figure 11-24 *Fluffy lies waiting to spring from his trap*

anyone will see them during its flight. The simple blob of fur design as shown in Figure 11-23 has been extremely effective at turning otherwise tough guys into quivering cowards during every launch, and has not suffered any damage from launching or landing.

Figure 11-24 shows what the entire system will look like when it is completed and loaded, waiting for me to lead my next victim into Fluffy's terrifying flight path. As you can see, the springs and launch pad look like part of some elaborate trap that has caught some black furry critter now huddling at the back of the cage as if afraid of us (yeah, right)! All I have to do is lower the light a bit and avoid a direct hit on the photosensor with my

flashlight as I bring my buddies in for a closer look at what the trap caught and say, "Whoa, it looks like that rat that came back out of the sewer, dude. Do you think it's still alive?" Now I shine the light into the rear of the cage so we can all get a good look at the thick messy black fur. "How am I going to get rid of it? Did it just move?!" Building up the suspense as you lead your victim into the perfect position right before launch always helps get the air ripe with fear! With your buddy standing in the prime place, stop moving and tell him to listen as you think the thing in the cage may be making a noise.

"Listen—did you hear it?" Just as your friend is concentrating on the non-existent noise you claim to have heard, flick the flashlight across the photocell to spring fluffy free from his bondage. "KaPow!"—the trap doors burst open with a loud bang as the lunch pad blasts Fluffy out of the cage. At this point, you buddies' eyes will be as wide as pie plates, realizing that a black furry thing is flying through the air directly at the space they presently occupy. Figure 11-25 shows Fluffy exploding from his cage just as the trap doors fly open in a spectacular display of noise and speed.

What happens next is always the best part of this prank! I have classified my victims into screechers, crouchers and jumpers, as these three responses are the most common, although it is often a combination of them. Screechers let out the kind of yelp that would normally be heard coming from a young girl who has just been introduced to a spider, which sounds extremely hilarious when coming from a person who claims to be extremely tough or just looks the part. Crouchers are silent but amazingly fast at the "duck and cover" maneuver, able to make the transition from the standing position to what looks like an awkward preying position with their hands over their head. Jumpers are by far the best victims as they find a way from the floor where they stand to the nearest bit of furniture at such a great speed that it appears as though they have found a way to travel faster than light. Often jumpers are also screechers, although the screech response is usually delayed until after they are a few feet off the ground safely

on top of the nearest table top. Sometimes Fluffy lands in a place that keeps him hidden from view, so the prank can last as long as you can keep a straight face or as long as the nearest table can support the weight of whoever may have jumped up on it for cover. As Fluffy leaves his cage (Figure 11-26), I'm sure you will discover many new types of responses from your victims. I have yet to see a person just stand calmly as if nothing is happening. Without a doubt, this project is one of the most effective Evil Genius pranks I have ever unleashed on to those whom "I owed one" to.

I hope you have fun with your flying rabid critter and many of the other evil pranks presented in this book, I know I have. Fluffy is great for parties, camp, or simply introducing new house guests to your flavor of Evil Genius humor, something that is often only appreciated by those with the knack to hack.

Congratulations—you have made it through 51 fun high-tech practical jokes that will keep you busy planning, plotting, and creating for quite some time. Don't be surprised, though, if your friends are skeptical about visiting you at your home once you've unleashed some of these pranks on them. They'll have good reason not to trust you! I think that designing and building your own

Figure 11-25 *Fluffy has had enough of this silly cage*

Figure 11-26 *Fluffy is on the war path, look out!*

gadgets, and watching the reactions of unsuspecting victims is highly entertaining, better than anything on TV these days. However, not everyone will appreciate your sense of humor. Testing your creations on people you don't know very well or anyone with a heart condition is definitely not a good idea. Have fun, but remember to use good judgment when selecting your practical joke victims. This book has been a lot of work, but also a lot of fun to put together. Hope that you have learned some new things along the way, and we look forward to hearing all about your own Evil Genius projects on the atomiczombie.com forum. Cheers, friends.

Chapter 12

Digital Fakery

Have you ever seen bizarre photos in tabloid magazines and on the Internet that look like they've been digitally enhanced to elicit shock and awe? You have probably wondered how they are created, since they appear to be real, but are obviously not. Well, when you are done reading this chapter, you will be able to manipulate pixels to create your own quality digital fakery. Armed with some basic computer software, a few digital photos and a mouse, you can become a "pixel ninja," putting yourself into scenes you have never been, altering your physical appearance, swapping your buddy's body parts, or simply messing with images in such a way as to shock and baffle viewers. Many of the same techniques used to manipulate still images can also be used to edit video, so you can have a lot of fun with just about any type of digital medium, even audio. Who knows, you may pull off a true work of art and become the creator of the next huge Internet legend, or simply get a co-worker back for a prank they played on you. On a more useful note, the skill of working with digital imaging software can land you a great career in movie production, advertising, or web design if you become a master of the art, something that takes only practice and a desire to be creative. So, let's begin your digital fakery lessons by doing wacky things to otherwise mundane photos.

Editing software

When it comes to working with still images, Adobe Photoshop is currently the king of the hill. Sure, there are many alternatives, some of them even free, but currently, Photoshop is a standard software program, and if you plan to pursue multimedia as a career, it makes sense to use the tools the professionals use. Because Photoshop is a high quality program, and many of the good alternative software programs are very similar to Photoshop, I will use it to demonstrate the technique of altering still images, which you can probably follow along with practically any image editing software. These simple tutorials are bare-bones, showing only a small fraction of what can be done to manipulate digital images. Of course, armed with only these basics and a lot of patience, you can alter photos in ways that are truly out of this world, or in subtle ways that even a trained eye could not detect the fakery. Many image editing programs are full of complex options, some of which you will find useful and some you may never touch, so the best way to learn is by actually using a tool or option to learn what it does. You could spend a lot of money on technical manuals about your favorite program, but the reality is that nothing beats a hands-on approach. It will cost you some time, so follow along with the instructions, then let your mouse run wild to see what you can create. At the time of writing this book, there were many alternatives to Photoshop, and one of the most popular is called Gimp, which is available for free for Windows, UNIX and Mac operating systems. If you do not have an image editing package and want to see what is currently available, just enter keywords such as "Photoshop Alternatives" or "Image Editing Software" in any Internet search engine. Your digital camera may have included some free imaging software, but beware, many times this software is a highly restricted version, limited to simple tasks like red eye removal or adding goofy frames to photos. Before you load your PC with megabytes of "bloat-ware," make sure the software does something useful.

Original photo quality

Let's get down to business. I assume that you have a digital camera and know how to get the photos onto your hard drive. Of course, you may simply want to alter photos that were sent to you by email or found on the 'Net, but either way, you will need to open them up in your photo editor for manipulation. The quality of the images you plan to mess with should be fairly high to start with. A 320 × 200 pixel image saved as a low quality jpeg will be difficult to manipulate, since it will be blocky from the compression and look like an obvious fake when you start adding clean pixels of effect to it. Sure, you can save a faked image with low compression to add to the realism and hide some of the telltale signs of manipulation when you are finished with it, but in the beginning you must have a decent quality image or images to work with. Photos from your digital camera should be initially saved in high or medium quality to preserve color information and resolution. If you are "borrowing" photos or clipart from the 'Net, then search for high resolution images with dimensions of 640 × 480 or greater (Google lets you choose image size when looking for images).

Download a simple portrait of your own beautiful mug from your digital camera, and we will start with some simple filters and effects that will transform your face in ways that Picasso could have never dreamed of. Figure 12-1 shows a simple mug shot with a minimal background, perfect for messing around with.

Open your mug shot in the photo editing software, and then place the image in a window that fills most of the screen by stretching the window handles. If your image is fresh from a camera or scanner, then it is probably scaled down to fit into the window, and will show a percentage value in the window title or under some image properties window. An image with a high resolution like 3200 × 2400 is much too large to email or place on a website. It is best that you do any manipulation, color correction or changes to the full-size image, and then save it as a smaller file afterwards as this will preserve maximum quality during the editing process.

Warping effects

Before doing any editing or scaling, I like to do simple hue, saturation, brightness and contrast

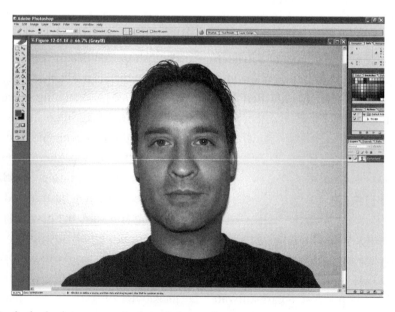

Figure 12-1   *I pity the fool who lets me manipulate their mug!*

Figure 12-2   *The Liquify filter window in Photoshop*

correction to a fresh image in case it needs it. Often a digital camera will vary the hue of the background or skin tones, or the scene may be dimly lit depending on the automatic exposure settings, so these simple corrections are usually an automatic response. With your full size mug shot open and ready for editing, click on the window to select it and then look for the liquify or warp filter on your toolbar or editing menu. In Photoshop, the liquify filter is found by clicking on "Filter" from the main menu and then by selecting "Liquify," which will bring up a new editing window as shown in Figure 12-2. Most image editing programs have an effect like liquify, although it may be called something else. There are even full programs that are specifically designed for this single task, like Kai's PhotoSoap, or Morpheus Photo Warper. To find these programs, search Google for "face warping software." The liquify effect is a very smooth organic method of pushing, pulling, shrinking, and expanding groups of pixels, which is why it is very powerful and can create results both believable and completely ridiculous depending on how it is used. Fashion magazines will often stretch the legs of models, fatten the lips, or warp other body parts in order to give a model a more perfect-looking body or face, although the effect is used sparingly so the result does not look manipulated.

In the liquify window, you will notice a toolbar with effect types on the left and a brush option pane on the right. From the left toolbar, select either the "Pucker" or "Bloat" tool, and then

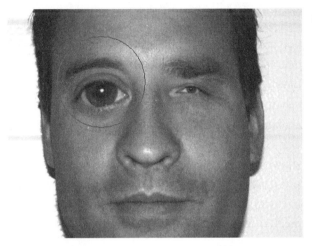

Figure 12-3   *Bloat (left) and pucker (right) applied to the eyes*

choose a brush size that is about twice the size of one of your subject's eyes. To find a new tool you do not recognize, hover over each tool until the text tip pops up indicating which tool you are pointed at. Figure 12-2 shows the brush as a dark circle which has grown to twice the size of my eyes as the size slider is slowly moved to the right. Now only the area directly inside the brush circle will be affected by whatever tool you have selected. Place the circular brush over one of the eyes so the pupil is in the center and then hit the left button on your mouse. As shown in Figure 12-3, the eyes magically shrink or expand depending on which tool (pucker or bloat) you choose. Depending on how long you hold down the button, the effects range from subtle and believable to almost creepy.

Press the "Restore All" button on the right to bring back the photo to its original state and then change the brush size and length of application to get the feel for this powerful shrink and expand toolset. If you want to keep certain areas from being effected, click on the "Freeze" or "Thaw" tool from the left toolbar and paint the area you want to mask out of the effect, this way you can expand or shrink using a large brush and exclude certain areas. An example of using the freeze tool may be when you want to shrink a large area around the eyes, yet keep the eyebrows in tact. Again, play around and get the feel for the tools, it is the only way to learn. When you are happy with your pucker and bloat alterations, click "OK" to return to the main interface and this will apply your changes. Now you can re-enter the liquify filter and make more changes that can be tested and reverted just like you did with the bloat and pucker tool. Figure 12-4 shows my results after bloating the eyes and nose and applying a little pucker on the lips. Also shown in Figure 12-4 is the best pixel mangling tool of all, the "Forward Warp Tool." Click on it!

This tool affects the area under the brush as if it were a magnet dragging the pixels around in a nice organic fashion. Again, the effect depends on the size of the brush and the areas you might have frozen, but the rules are the same; the larger the

brush, the more pixels you will alter at once. As you will soon see, this tool is the ultimate weapon against normality, with the ability to alter photos in such a way that you can change a person's facial structure in seconds from something different yet real looking, to the most outlandish goofy cartoon figure imaginable. I don't care how old and "mature" you are, it is impossible not to giggle as you alter your buddy's poker-face mug into a living cartoon. A little pull here, a push there, and before you know it, the face will look even worse than mine does in Figure 12-5, although I do think the hair is working for me!

Have fun, and play around with these three liquify tools and you will find out that you can do a lot with the proper application of one or more of those tools. If you are working with photos that include a complex background, then you may have noticed that large brushes will affect the area around the subject as well, which can make the image look obviously faked. To get around this, freeze the area around your subject before applying the effect and the background will stay mostly unaffected. Try the liquify tools on non living things like cars or houses as well, and you will see that you now have the power to alter reality in any way you like.

Making hoax photos

The art of compositing images to make a fake has been used since the invention of the camera, and if done properly, the result can be extremely believable, enough so to start an urban legend (evil gears turning). UFOs, Big Foot, ghosts, fairies, and many other ideas come to mind, and because of the powerful digital editing software available today, you can pull off these hoaxes without spending days in a dark room doctoring photos with smoke, mirrors and a paint brush. Ever since I was abducted by aliens and implanted with a mind enhancing computer, I have been fascinated by images of UFOs and the people who believe they are real.

Figure 12-4   *Getting ready for round two with the warp tool*

Figure 12-5 *Dude, that's a mullet to be proud of!*

I will admit to making a few really good UFO photos years back, which are to this day circulating on certain websites with captions like "is this real," "we are looking for the photographer," and so on. Why would I do such a thing you ask? Well, for the same reason you are reading this text, because we think it's funny!

Let's mock up a simple UFO sighting using household items and some of the basic editing tools that almost all image editing programs have. The amount of effort you put into your images will directly reflect the result, and if you plan the shots, keeping obvious signs of fakery hidden, you can create an image that only a true forensics expert could detect. Shadows, lighting, pixel edges, reflections, focal length, and color are all things to be mindful of as you composite images, but with today's powerful graphics editing packages, many of these things are easily corrected.

Let's begin our UFO encounter with a boring shot of some dude standing in the front yard pointing towards the sky as if signaling beings from another world to take him away (Figure 12-6). A perfect UFO fake should look impromptu, as if the photographer just happened to be snapping photos, or had to scramble to get the camera and take the shot. An area that cannot be identified is also good, since there will be no other witnesses. Placing a giant cigar shaped object over Times Square at noon and trying to pass it off as real would be a waste of time, since there would have been thousands of eye witnesses to the "invasion," yet you are the only one who happened to notice it.

When taking the initial shot, try to imagine where you plan to insert the UFO so you can leave enough room for it. Again, the larger the image the better, since you will have a large clean area to start with and can later save a cropped version with lower quality to specifically add more "foo" to the fight. I wanted a huge saucer shaped chrome mother ship to appear directly over the house, casting a large ominous shadow as it approached, so a large area of the sky was included in the photograph. With your base photo ready for compositing, you can now create the actual UFO. There are many ways to do this, and I have used three methods with great success. One of the best ways to fake a UFO is to set up your initial shot on a tripod, photographing the background first and then taking a second shot from the exact same location of an actual fake UFO as it is flung into the air or hung by a very thin wire. The two images are then pasted together with a little brushing around the UFO to hide the edge pixels,

Figure 12-6   *The start of our UFO composite*

resulting in a fake that can look exceptionally real depending on the quality of your fake UFO model. The obvious benefits to this method are the fact that the lighting, shadows, reflections, and colors will be perfect for both the background and the UFO. The problem is making the UFO large enough to look menacing, and then finding a way to get it in the air for the photograph. Throwing a silver colored lamp base up in the air to snap a shot is just not going to cut it; you need something more alien looking and much larger. Find smooth organic object to model your craft, objects that are not obviously recognizable. Add strange lights and accessories to the craft so the "experts" can theorize how the antigravity engine works. "Hey, it looks like a David Hamel antigravity rotator with a Bedini Overunity Generator." Yes, do a good fake, and all the experts will have an opinion! Another option is to render your UFO in a powerful 3D program with radiosity ray tracing directly over the background. This technique requires a mastery of some very complex software such as Maya or 3DS Max, but can deliver some very stunning results if you know what you are doing. Most movies are

done this way, and as you can see, often they look pretty realistic. The final option I have used quite often is to create a smaller, more complex UFO model and photograph it separately, compositing it into the background photo. This approach has the benefit of not requiring a full scale mother ship to be hung or thrown in mid air, so you can make a very good small scale model. The downside is that the colors, shadows, reflections and field of view all have to be carefully matched between the model and the background, or people like "The Lone Gunmen" will figure out your fakery. Let's use the later method to create the UFO, as I don't think my neighbors will appreciate yet another 16 foot mother craft hanging over their yard on a wire. To be silly, I stuck an LED flashlight to the base of our dog's stainless steel water bowl, held it up to the ceiling and snapped a photo. Figure 12-7 shows how the bowl looks like a classic low budget film flying saucer, complete with photonic thrusters (LEDs).

Sure, this model is lame, but the idea here is show the compositing techniques, not the model, which I will leave to your twisted imagination.

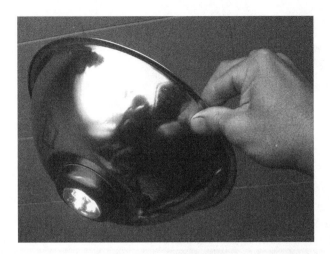

Figure 12-7   *The classic flying saucer*

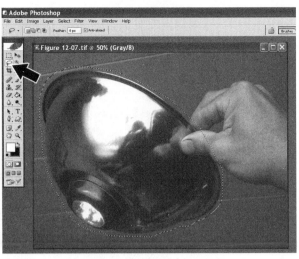

Figure 12-8   *Cutting out the UFO using the lasso tool*

Another method that could be used here to increase the realism of the model would be to photograph it outdoors in cloud cover that will roughly match the original scene shot, this way overhead reflections and contrast will match. To take that a step further, print out a color page of the original background shot and have a helper hold it near the chrome UFO body so it reflects, as if actually in the scene. You could even take a separate "scene reflection" shot that would better approximate the true angle of reflection that would be seen on the UFO just to beat the forensic experts who would pick over your photo under 500x magnifications looking for anomalies. My UFO shot is about as bad as you can get, and it even includes my thumb, but the final photo will still look somewhat real, although completely laughable. Now, it's mixing time. Open your UFO shot and then use the lasso tool to select it by drawing a border around the entire UFO as shown in Figure 12-8. Do not try to lasso the edge of the object you are compositing or it will look truly fake due to the sharp edge transitions and imperfect curves. As shown, I have included a small bit of the background around the UFO body in the lassoed area. There are filters and programs specifically designed for background removal, but unless the background is perfect, you are going to get the best result by hand erasing the background around the edges of the object. The arrow is

pointing the lasso tool in Photoshop, and since this is a basic tool included in any photo editing software, it will most likely be visible on the toolbar. When you let go of the left mouse button, the lasso closes and you now have the object selected. If you made a mistake, click on "Select" and then "Deselect" from the menu or just draw another lasso. With the UFO lassoed, click on "Edit" and then "Copy" from the menu, and it will put all the pixels inside the lasso into the memory (clipboard). Copy and paste are also common functions that all image editing programs will have, and often they have keyboard shortcuts.

Now minimize or close the UFO model photo and select or open the background photo for editing. Since the UFO is now in memory, you can go to the "Edit" menu and select "Paste" to drop it right over your background photo. The UFO will become a "layer" in your image and at this point will be too large, in the wrong position, and have extremely fake looking borders. Make sure "Show Bounding Box" is checked off in Photoshop so you can work with the UFO image as an easily resizable layer as shown in Figure 12-9. The points of the bounding box allow you to resize the UFO to better suit the scene. To move the entire UFO, click on the center point and drag it with the mouse.

Manipulate the UFO layer so it is placed and resized in the scene in whatever manner you think looks best.

Figure 12-9   *Using the bounding box to move and resize the UFO*

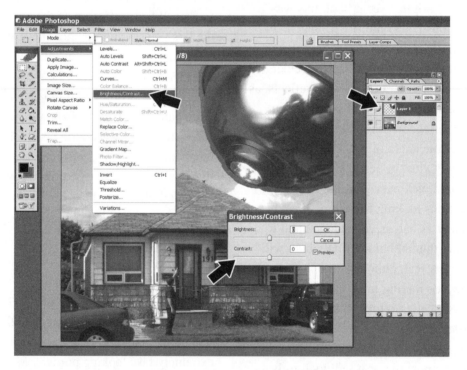

Figure 12-10   *Try to match hue, contrast and brightness on all layers*

You will notice that you can alter the aspect ratio of the object by moving the side points, making the UFO longer or flatter. I decided to make my UFO look a little more round, and then I shrunk it down and moved it into the top corner of the scene so it would look like I was pointing directly at it.

With the UFO where you want it, you can now try to match the color hue, brightness and contrast to best match the UFO to the background, which is necessary since it was photographed in different lighting. As shown in Figure 12-10, you make sure the UFO layer is selected from the "Layers" toolbar (far right) before applying effects to the UFO. Since the UFO was pasted over the scene, there is a background and UFO layer (Layer 1) created, which allows you to work on each image as a separate entity. Layers are very powerful, so get to know them if you want to increase the quality of your "Photoshop KungFoo." With the UFO layer selected, choose "Image," "Adjustments," and then "Brightness/Contrast" from the menu as shown in Figure 12-10. This will pop up the adjustment sliders. The same should be done with "Hue/Saturation."

Slight changes of brightness and contrast should be made, followed by changes in hue and saturation, and then the cycle should be repeated. This fine tuning of layers will get your UFO looking like it was photographed in the same light as the background, so have patience. Yes, you could also select the background layer and adjust its levels as well. Ignore the horrible edges and try to make the UFO look like it actually belongs in the scene. Once you are happy with the result, choose "View" and then "Zoom In" a few times until you have a large view of the edges of the UFO as shown in Figure 12-11. Edge removal is fine work, so 300x or higher magnification should be used if you want the results to be believable.

Remember, your work my come under high magnification when the cynics try to disprove it, so

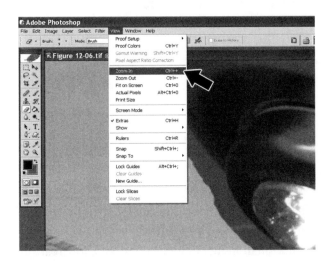

Figure 12-11 *Zooming in for some fine detailed work*

try to stay one step ahead of them! When you are zoomed in this much, you will have to move the window sliders to center the area you plan to work on, so pick a place to start by centering it in the window. We will now erase the background edges manually using the eraser tool. This may seem like a ton of work, but it really does not take all that long, especially on a simple shape like this. Choose the "Eraser Tool" from the tool menu as shown in Figure 12-12, and then click on the little down arrow next to the brush top open the brush panel. As you can see, there are many brush options and sizes to choose from, so try to find one that is of "medium edge softness" and small enough to work with at such a high zoom level. If you are new at this, then try experimenting with the various brush type and sizes until you find one that lets you rub out the edges without leaving an obvious jagged edge or uneven line. I like a smooth eraser brush with a diameter of 8-10 pixels for edge removal, especially at such a high zoom level.

Since the UFO is a layer overtop of the background, the erase essentially removes pixels from the UFO image as if you were rubbing them out. Take your time and try to make smooth, long strokes around curves so they do not look jagged or bumpy. It takes a bit of practice to get the feel for "mouse painting," but with practice, you will be able to trace a border with deadly precision. As shown in Figure 12-13, I am using a very small soft brush (black circle) to erase the edges of the UFO by carefully tracing the border. If you make a mistake, you can always use the "Edit," "Step Backwards" function to undo your last changes.

Carefully erase the entire border around the UFO by centering the work area in the window a bit at a time. If you cut into the UFO with the erase, use the undo function and try again. When you are finished, unzoom the scene to inspect your work. With time, you will get better at edge tracing and find that this method of background removal is better than any

Figure 12-12   *Selecting an appropriate eraser brush*

Figure 12-13   *Carefully eraser brush the UFO edges*

other, including the chroma keying method, which attempts to do the removal automatically by erasing a single color from the background. Figure 12-14 shows the final UFO without any border, hovering above our house right before abducting me for another interstellar journey. You know, it drives me crazy having to advance the time on my watch by one hour every time my alien friends visit!

Adding reflections and shadows

The largest telltale signs of fakery are reflections and shadows, which are often missing. Working with a small model like this does not allow real shadows to be cast in the scene, so you are going

Figure 12-14   *No more edges. The compositing continues*

Figure 12-15   *Duplicating the UFO layer to make a shadow*

to have to fudge it using the editing software. With a little knowledge of light and some basic tools, a very convincing shadow can be made using almost any editing program, so let's dive right in. Trying to use a darkening brush to hand paint the shadows in will likely result in a worse composite than one with no shadows at all, so let me show you a little trick I use to add shadows into images. We will

make a copy of the UFO layer and turn it into a shadow, so the shadow looks like it was actually cast by the image, which will fool 99 percent of those who are putting your photo under scrutiny. As shown in Figure 12-15, you must first select the UFO layer by clicking on "Layer 1" on the layer panel. Now click on "Select" and "All" to lasso the entire scene. Now you can select "Edit," "Copy," then "Edit" and "Paste" to make a duplicate of the UFO layer. The second layer will be called "Layer 2" and most likely be dropped in the center of your screen.

With the new UFO layer selected, choose "Edit," "Transform" and the "Distort" as shown in Figure 12-16, which will allow you to drag the bounding box points independently, creating a distorted image, much like the way a showdown would be unevenly cast from an object onto a surface. You don't have to be a ray tracing expert to make this effect work, just try to imagine how the shadow would fall on whatever object you are adding it to. As long as the shadow does not look too perfect, or seems to flow against other shadows in the scene, it will probably fool anyone who looks at the image.

As shown in Figure 12-17, I distorted the shadow layer and moved it over the roof of the house, imagining that this is how it would look if cast by the UFO blocking the sun. When you are done distorting the bounding box, click on the move tool as shown in Figure 12-17 to apply the transformation, which will permanently distort the layer. At this point, the layer does not look like a

shadow, just a twisted UFO, so we need to apply a few more common filters to fix it.

A real shadow does not block all light, so it does not blank out all the features of the object which it is cast, so we need to add opacity to our layer to allow some of the original background to show through. To do this, make sure the shadow layer is still selected, then choose "Layer," "Layer Style" and then "Blanding Options" to bring up the very powerful and often used Layer Style panel, which will allow you to adjust how one layer interacts with another. This is shown in Figure 12-18.

Layers and layers styles are what give these programs their true power as you will see by using them over time. In the real world, colors and light interact with each other, so transparency and blending is key to making your work look convincing. The Layer Style panel is shown in Figure 12-19, and I have reduced the layers opacity to 50 percent, allowing a 50-50 bland between the shadow layer and the roof of the house. There are still a few steps needed to turn the new UFO layer into a shadow, but as you can see the ability to alter opacity is extremely powerful and can be used to make shadows and reflections very easily.

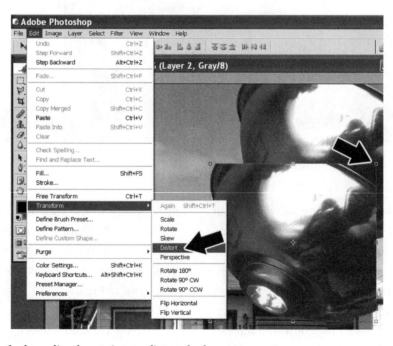

Figure 12-16 *Using the bounding box points to distort the layer*

Figure 12-17   *The distorted UFO layer soon to become a shadow*

Figure 12-18   *Getting to the Layer Style panel for a selected layer*

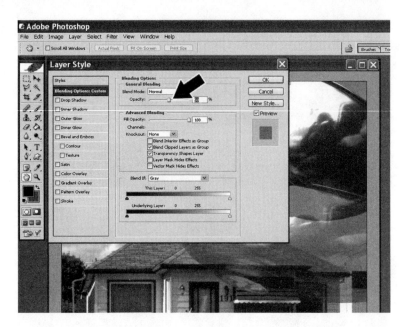

Figure 12-19   *A 50 percent blend with the new layer and the background*

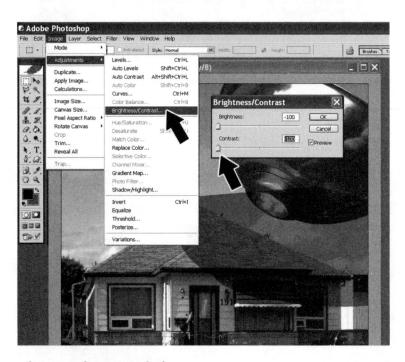

Figure 12-20   *The new layer now becomes a shadow*

With the shadow layer's opacity set to a level that looks good in your scene, choose "Image," "Adjustments," then "Brightness/Contrast" to remove all of the original image, leaving only a dark shadow on the background. As shown in Figure 12-20, the new layer now looks like a sharp cast shadow once the brightness and contrast are reduced to zero (-100). The shadow is still too sharp looking, and extends beyond the rooftop, but that will be easy to fix.

Due to the nature of the source lighting in a scene, some shadows are extremely sharp, while others are very blunt, so we must adjust our fake shadow to match other scene shadows. The newly created shadow is perfectly sharp at the edges since it was cut from an actual layer, so it will need some fuzziness around the edges in order to look real. Look at the other shadows in the scene to determine how sharp the edges of your fake shadow need to be in order to blend in. Now choose "Filter," "Blur," and then "Gaussian Blur" to bring up the Gaussian blur panel, which will do an amazing job of softening up the edges of your fake shadow. Gaussian blur is extremely powerful, and available in most image editing packages, and you will find it among one of the most widely used tools, especially when compositing or drawing cartoons. As shown in Figure 12-21, a little bit of Gaussian blur does wonders to the new shadow layer, making it look extremely realistic. Painting a similar shadow by hand would have taken a lot more time and eye–hand coordination.

When you have the blur set the way you like it, press "OK" to apply the transformation. Now you can go back and tweak the opacity if you think it is needed in order to make the new shadow look

exactly like the other shadows in the scene. With the proper opacity and Gaussian blur, the shadow will look like the real deal. So far so good, but the shadow seems to extend past the roof of the house into thin air, so it needs a little trimming. No big deal, just use the polygon or freehand lasso tool to select the areas of the shadow you do not want, and then select "Edit" and "Cut" to trim them away. As shown in Figure 12-22, the lasso can be used to cut or copy certain areas of a layer, which is why it is such a useful tool.

Now the shadow looks good, the UFO is pasted in the scene with no obvious jags around the edges, brightness and contrast looks good on all layers, but there is one problem that even a newbie would notice; the reflections in the bowl! Yes indeed, the reflection of a picture frame from the living room and my own head are tale signs of fakery and must be fixed before I can send this photo to the FBI. Another great tool is the Healing Brush, which will sample one area of an image and either directly replace it with another, or fudge the

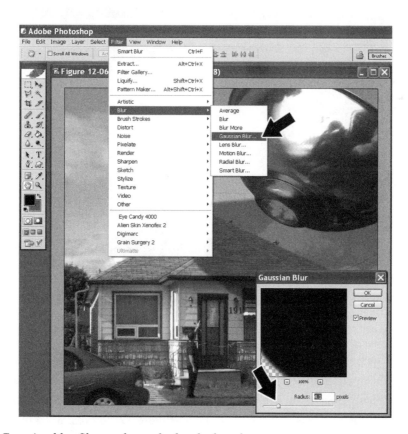

Figure 12-21   *The Gaussian blur filter makes a shadow look real*

Figure 12-22   *Lasso areas of the shadow to be trimmed*

Figure 12-23   *Selecting and using the healing brush*

two together to make a smoother transition. Besides photo fakery, the healing brush is great for removing litter from a photo, erasing unwanted graffiti, or simply hiding small objects that take away from a photo. The healing brush will do a

great job of fudging out the reflections in the dog's bowl, I mean UFO. Make sure the UFO layer is selected, then click on the healing brush as shown in Figure 12-23 (yes, it's a band-aid). Now choose a brush size by clicking the little down arrow and then select "OK." To heal an area, you must first find a similar area as a sample area and then press the "Alt" key with your mouse over that area. The area under the mouse is now sampled and you can press the left mouse button to heal the area as if you were painting. It takes a little messing around to get the area perfectly healed, but play around, use the undo function and you will eventually get the results you want.

I also used the healing brush to get rid of the logo on the LED light that was stuck to the bottom of the bowl UFO. Of course, who's to say that aliens would not be proud of their brand? Wouldn't NASA proudly display their logo on the hull of an interstellar spacecraft? Anyhow, let's finish up this grand fakery with a few more simple filters to add icing to the cake. The light cast from the photonic space drive seems a little weak, especially since none of it is being reflective onto the scene. Most editing programs have many light casting filters and effects, but one of the oldest and most effective is the classic lens flare. This simple filter simulates the flare that a multi-element coated lens from a quality set of optics may give when presented with a bright light source. I find this effect can also make cool fire balls, jet blasts, and can be used to wash out the lighting in a fake image, making it harder to detect bad edges or mismatched contrasts. Choose "Filter," "Render" and then "Lens Flare" to bring up the effect panel. As shown in Figure 12-24, the effect has a few basic settings like lens type, brightness and position, so just play around until you make it look good.

When you hit "OK," the lens flare will be applied to the entire scene, which makes it look like the light from the UFO engine is being cast into the scene and on the top of the roof. Sure, the floating dog dish UFO scene (Figure 12-25) is far from perfect, but the composite is pretty good. Shadows, color, brightness, contrast, lighting and edges are not a dead giveaway, although I can't say the same for the alien mother ship! Imagine what you could do with a little planning and some junk laying around the house? Yes, the next urban legend could be yours!

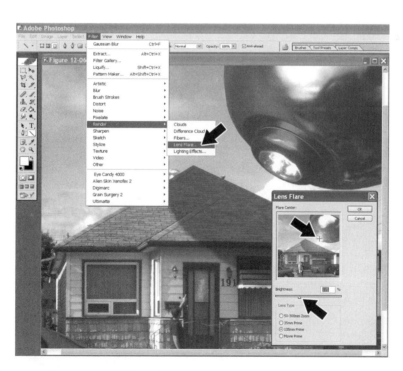

Figure 12-24 *Adding some heat to the UFO engine*

Figure 12-25 *Take me back to planet Lacidar!*

I wish there was a lot more room in this book to explore the thousands of amazing feature of Photoshop and many of the other image editing programs, but that would require another 1,000 pages, so I will leave you to your own evil devices to twist, distort and manipulate innocent pixels into works of deviant art. Like all skills, a hands-on approach is the best way to learn, so start clicking buttons with reckless abandon, who knows what you might discover!

Index

Headings in *italics* indicate projects (pranks).
Page numbers in *italics* indicate illustrations.

A

AC transformers 115–16
activator, electromechanical 155
amplifier, audio 61, 91–3
Audio clips 48–50
Audio signal distorter 59–61
   assembly *60*
   schematic *60*

B

Balance tester 115–17
   schematic *116*
ball switch 118–20
Bat, flying 165–70
   creating the beast 167–8
Beep Beep 20–4
   assembly *23*
   schematic *22*
Brain, alive and pickled 172–6
   schematic *175*
breadboard 14–15
Breathing creature 33–5
   assembly *35*
   schematic *34*

C

Cadmium Sulfide cells *see* CDS cells
cameras, disposable 120–4
capacitors 7–8, 23, 26–7
   flash capacitors 120–4
   value chart *9*
Car engine troubles 68–70
   assembly *69*
CD-ROM drive 136–42
CDS cells 19–20, 38, 50, 83, 85, 196–7
ceramic magnets 83, 139
circuit boards 4, 12–15
clanking hard drive 63–7
clock oscillator 57
coil winding 54–6, 82
Coin minting machine 136–42
   wiring diagram *141*
crawling creature *see Creepy Crawleys*
creatures
   breathing 33–5
   crawling 40–3
   jumping 44–7
   swinging 36–40
   trapped 198–202
Creepy Crawleys 40–3

D

DC motors 68–70
decade counter 66
Digital fakes 203–20
diodes 8–10 *see also* LEDs
distorting sound 59–61
Dog talk 149–53
Doll robot 95–8
   assembly *97*
drill, electric 165–6, 167–9
Dripping sound 17–20
   assembly *19*
   schematic *18*

E

electric shock 113–15, 115–17, 117–20, 120–4
Electric zapper 120–4
   assembly *123*
electromagnet 65–6, 82–4 *see also* solenoids
embedding messages
   audio 157–9
   video 160–4
Eyes in the dark 71–6
   assembly *75*
   schematic *75*

F

ferrite bead 55–6
FM transmitter 54–7
Footsteps at night 84–6
   assembly *86*
   schematic *85*
frequency adjusting 51–3, 56

G

Gas leak 124–7
   assembly *126*
   schematic *126*
Geiger counter *see Radiation detector*
ghost mirror 170–2

H

Hard drive, clanking 63–7
   assembly *67*
   schematic *67*
hard drive, dismantling of 64–5
headphones 91–3
healing brush 217–19
hoax photos 206–20

I

image editing software
   healing brush 217–19
   lasso 209
   light casting filter 219
   liquify filter 205–6
   shadows 212–17
induction shocker 115–20
   schematic *116*
infrared
   LEDs 51–3, 91–4
   transmitter and receiver 92–5

J

Jack-O-Lantern *see Pumpkin lantern alive*
jumping creature 44–7

K

Knock Knock 24–7

L

lasso 209
launch pads 44–5, 188–90, 194–5
Launcher, universal 44–7
law 54, 63
LEDs 10, 71–2, 74–5, 86–8, 133–6, 147–8, 172–6
   infrared 51–3, 91–4
   sequencer schematic *175*
   wiring schematic 87, *148*
Lie detector 147–9
   assembly *148*
   schematic *148*
light-activated trigger 38, 44–7, 187, 193–8
   assembly *197*
   schematic *39, 46, 196*
Light bulb, magic 133–6
   dismantling 133–4, *134, 135*
   wiring 135, *135*
light casting filter 219
light emitting diodes *see* LEDs
light intensity 86–7
light-sensitive resistors *see* CDS cells
light switcher 104–7
lines 41, 165, 167–9
liquify filter 205–6

M

magnets
   ceramic 83, 139
   electromagnet 65–6, 82–4
   NIB (neodymium) 83, 107–10
   permanent 82–4
MCD (millicandela) 86–7
mercury switch 118–20
microphones 76–9
microswitches 48–50, 156
millicandela (MCD) 86–7
Mind control, audio 157–9
Mind control, video 160–4
Mirror, haunted 170–2
motion-activated switch 118–20
motors 40–2, 68–70, 74–5, 84–6, 137–8, 140, 154–5
   see also servo
   schematic *42*
   windshield wiper 176–9, 183–4
Mover, universal 176–9, 183–4
moving objects 176–9
multi-meter 4

N

neodymium magnet 107–10
NIB *see* neodymium magnet
Noise nightmare 76–81
   recording 77–9
   task scheduling 79–81
noises *see* sounds
NPN *see* transistors

O

ohms 5
oscillators 21, 57, 127, 130
Ouija board, animated 107–11
   magnet installation 109

P

paper-maché 181–3
permanent magnets
phone jack, rigged 62
photo hoax 206–20
photocells *see* CDS cells
Photoshop, Adobe 203
piezo elements 17–20, 23–4, 113–15, 125, 127
ping-pong balls 71–3, 183–4
PNP *see* transistors
polygraph *see* Lie detector
Pop-up head 101–4
   launching mechanism 103
potentiometer (pot) *see* resistors, variable
pulleys 165, 169
Pulse checker 113–15
Pumpkin lantern alive 181–5

Index

R

Radiation detector 127–31
   assembly *130*
   schematic *130*
radio controlled
   car 154–5
   servo 95–8, 105
Radio station blocker 54–7
   assembly *56*
   schematic *55*
radio transmitter 153–6
Rats and mice 82–4
   assembly *84*
   schematic *83*
RC *see* radio controlled
receiver 91–5, 149–53
   schematic *94*
recording and playback 48–50, 77–9
remote control
   infrared 51–3, 91–3
   radio controlled 95–8
Remote control jammer 51–3
   assembly *53*
   schematic *52*
remote switch 118–20
resistors 5–7
   color band chart *6*
   light-sensitive *see* CDS cells
   variable 5–6, 52–3, 61–2, 130, 196

S

semiconductors 2 *see also* capacitors; diodes; resistors;
   transistors
servo, radio controlled 95–8, 105
Shadow projector 86–9
shadows 212–17
Shocking box 117–20
   assembly *119*
   schematic *118*
Signs from beyond 104–7
smell, foul *see Stink generator*
soldering 2–4
solenoids 24–7, 37, 194–6
sound
   distorting 59–61
   editing 157–8
   recording 48–50, 77–9
Sound-activated switch 179–81
   assembly *180*
   schematic *179*
sounds *see also* recording and playback
   beeping 20–4
   car engine troubles 68–70
   clanking hard drive 63–7
   dripping 17–20
   knocking 24–7, 104–7
   nightmare 76–81

speakers 20–4
spider launcher 36–40
spirit board *see Ouija board, animated*
springs 44–5, 103, 188–90
Stink generator 28–32
   assembly *31*
   schematic *30*
   stink formula 29–30
subliminal control *see Mind control*
swinging creature 36–40
switches
   mercury 118–20
   microswitches 48–50, 156
   motion-activated 118–20
   remote 176–9
   sound-activated 179–81

T

Telepathy 153–6
telephone audio interface 99–101
   schematic *100*
Telephone jammer 61–3
   assembly *63*
   schematic *62*
timers (555) 17–19, 20–4, 24, 26–7, 66, 74–5, 82–3,
   85–6
tools
   multi-meter 4
   soldering 2–4
transformers, AC 115–16, 118, 119
transformers, audio 100–1
transistors 10–12, 38–9, 92–4, 196–7
transmitter 91–5, 149–53
   schematic *92*
Transparent walls 142–6
Trap door cage 190–3
Trapped creature 187–202
trigger, light-activated 38
   schematic *39*
TV-screen 170–2

U

UFO photo hoax 206–20

V

video
   clips 144–6
   editing 160–4
video camera 142–6, 170–2
Video signal distorter 57–9
   assembly *59*
   schematic *58*
Voice changer 98–101
   schematic *100*
voice coil 65–6

Voices from beyond 91–5

walkie talkie 149–53
wall knocker 104–7

white noise generator 124–7
   assembly *126*
   schematic *126*
windshield wiper 176–9, 183–4

W